Lecture Notes in Economics and Mathematical Systems

609

Founding Editors:

M. Beckmann
H.P. Künzi

Managing Editors:

Prof. Dr. G. Fandel
Fachbereich Wirtschaftswissenschaften
Fernuniversität Hagen
Feithstr. 140/AVZ II, 58084 Hagen, Germany

Prof. Dr. W. Trockel
Institut für Mathematische Wirtschaftsforschung (IMW)
Universität Bielefeld
Universitätsstr. 25, 33615 Bielefeld, Germany

Editorial Board:

A. Basile, A. Drexl, H. Dawid, K. Inderfurth, W. Kürsten

Yong Fang • Kin Keung Lai
Shouyang Wang

Fuzzy Portfolio Optimization

Theory and Methods

 Springer

Doctor Yong Fang
Academy of Mathematics and Systems Science
Chinese Academy of Sciences
Beijing 100080
China
yfang@amss.ac.cn

Professor Kin Keung Lai
Department of Management Sciences
City University of Hong Kong
Tat Chee Avenue
Kowloon, Hong Kong
mskklai@cityu.edu.hk

Professor Shouyang Wang
Academy of Mathematics and Systems Science
Chinese Academy of Sciences
Beijing 100080
China
sywang@amss.ac.cn

ISBN 978-3-540-77925-4   e-ISBN 978-3-540-77926-1

DOI 10.1007/978-3-540-77926-1

Lecture Notes in Economics and Mathematical Systems ISSN 0075-8442

Library of Congress Control Number: 2008925545

Cover design: WMX Design GmbH, Heidelberg

Printed on acid-free paper

9 8 7 6 5 4 3 2 1

springer.com

Preface

Most of the existing portfolio selection models are based on the probability theory. Though they often deal with the uncertainty via probabilistic approaches, we have to mention that the probabilistic approaches only partly capture the reality. Some other techniques have also been applied to handle the uncertainty of the financial markets, for instance, the fuzzy set theory [Zadeh (1965)]. In reality, many events with fuzziness are characterized by probabilistic approaches, although they are not random events. The fuzzy set theory has been widely used to solve many practical problems, including financial risk management. By using fuzzy mathematical approaches, quantitative analysis, qualitative analysis, the experts' knowledge and the investors' subjective opinions can be better integrated into a portfolio selection model.

The contents of this book mainly comprise of the authors' research results for fuzzy portfolio selection problems in recent years. In addition, in the book, the authors will also introduce some other important progress in the field of fuzzy portfolio optimization. Some fundamental issues and problems of portfolio selection have been studied systematically and extensively by the authors to apply fuzzy systems theory and optimization methods. A new framework for investment analysis is presented in this book. A series of portfolio selection models are given and some of them might be more efficient for practical applications. Some application examples are given to illustrate these models by using real data from the Chinese securities markets. The main innovative results of this book include: portfolio selection models with fuzzy liquidity constraints in a frictional securities market are proposed; based on the fuzzy decision theory, fuzzy portfolio selection models with S shape fuzzy numbers are formulated; an estimation approach for interval returns of securities is proposed; the concept of semi-absolute deviation interval risk function is given, portfolio selection models with interval returns and interval risk are formulated; and the semi-definite programming approach for estimating possibility distribution of returns of securities is proposed. Moveover, the center spread possibility distribution portfolio selection models in a frictional securities market are formulated, and the four fuzzy index tracking portfolio selection models are formulated, based on the four different measuring methods for tracking index error.

We would like to thank many friends and colleagues for their help and support in preparing this monograph. First, we thank Prof. Shushang Zhu of Fudan University, Prof. Jiuping Xu of Sichuan University, Prof. Masao Fukushima of Kyoto University and Prof. Duan Li of Chinese University of Hong Kong for their contributions to the book. Three chapters are based on the results that we achieved jointly with them. We would like to thank several scientists for their helpful suggestions and valuable comments on our research in this area, among them are Prof. Shu-Cherng Fang of North Carolina State University, Prof. Baoding Liu of Tsinghua University, Prof. M. Makowski of International Institute of Applied Systems Analysis, Prof. Yong Shi of University of Nebraska at Omaha, and Prof. Y. Yamamoto of Tsukuba University. Besides, we have to thank many colleagues who made important contributions in this promising area, including Prof. Hideo Tanaka of Hiroshima International University, Prof. Peijun Guo of Kagawa University, Prof. Masahiro Inuiguchi of Osaka University, Prof. Srichander Ramaswamy of Bank for International Settlements, Prof. Enriqueta Vercher of Universitat de València and Prof. Christer Carlsson of Åbo Akademi University because their research stimulated us to join this area of research. Finally, we would like to thank the National Natural Science Foundation of China, Chinese Academy of Sciences (CAS), Academy of Mathematics and Systems Science of CAS, Hong Kong Research Granting Committee and City University of Hong Kong for their financial support to our research.

Yong Fang
Institute of Systems Science
Academy of Mathematics and Systems Science
Chinese Academy of Sciences
Beijing 100080, China
Email: yfang@amss.ac.cn

Kin Keung Lai
Department of Management Sciences
City University of Hong Kong
Tat Chee Avenue, Kowloon Hong Kong
Email: mskklai@cityu.edu.hk

Shouyang Wang
Institute of Systems Science
Academy of Mathematics and Systems Science
Chinese Academy of Sciences
Beijing 100080, China
Email: sywang@amss.ac.cn

Contents

Part I

Literature Review

1

Survey for Portfolio Selection Under Fuzzy Uncertain Circumstances

1.1 Introduction

Uncertainties about future events make the behavior of economic indicators unpredictable and, at times, brings about turbulence to financial markets. Assumptions about their behavior, while allocating resources, under an uncertain and ever-changing environment, are the building blocks for theories of economics and finance. The theories have been used to apply mathematical analytical tools to model both the behavior of the economic agents, and future events in financial markets. Resource allocation methods derived from modern mathematical models, in turn, play an influential role in work practices of financial institutions, and in a way, become a not-insignificant tool used in the financial markets.

It has been suggested that the origin of modern mathematical models in finance can be traced back to Louis Bachelier's dissertation on the theory of speculation. However, without doubt, the ground-breaking work of Markowitz (1952) in portfolio selection has been the most impact-making development in modern mathematical finance management. The Markowitz theory of portfolio management deals with individual agents in the financial markets. It combines probability and optimization theories to model the behavior of agents of economic change. The agents are assumed to strike a balance between maximizing the return and minimizing the risk of investment decisions. Return is quantified as the mean, and risk as the variance, of the portfolio of securities. These mathematical representations of return and risk have allowed optimization tools to be applied to studies of portfolio management. The twin objectives of investors - profit maximization and risk minimization - are thus quantified, so as to maximize the expected value and to minimize the variance of the portfolio value. The exact solution will depend on the level of risk (in comparison with the rate of return) they would bear. Even though many later models have different views on mathematical definitions of risk and return of economic agents, the trade-off between return and risk has always been the major problem for the theories to solve.

Although we have seen that the basic portfolio analysis problem in the mean-variance model can be solved in polynomial time, historically, major efforts have been made to reduce computational requirements. The single index model of Sharpe (1963) is an early breakthrough in this direction - it reduces the estimation of $O(N^2)$ variance-covariance coefficients to a total of $O(N)$ parameters. Based on the observation that investors may only be concerned with the risk (of return) being lower than the mean (downside risk), the Mean-Semivariance (E-S) method was proposed to design the model (see, e.g., Markowitz (1959), Mao (1970) and Swalm(1966). Semivariance is defined as the expected value of squared "positive(or negative)" deviations from the mean (or more generally, a value chosen by the decision-maker as a critical value).

In order to solve large-scale portfolio optimization problems, Konno and Yamazaki (1991) considered mean-absolute deviation as the risk of portfolio investment. Using the historical data of Tokyo Stock Exchange, Konno and Yamazaki compared the performance of the Mean Variance Model and the Mean Absolute Deviation Model and found that the performance of the two models was very similar. Feinstein and Thapa presented a reformulation of the MAD model, which is equivalent to the model of Konno and Yamazaki, and at the same time, reduces the bound on the number of non-zero assets in the optimal portfolio by half. While Konno and Yamazaki showed that the mean absolute deviation model did not require the covariance matrix, Simaan (1997) found that this would result in greater estimation risk, which outweighed the benefits.

The third moment of a return distribution is called skewness, which measures the asymmetry of the probability distribution. A natural extension of the mean-variance model is to add the skewness as a factor for consideration in portfolio management. There will be three goals: maximizing mean and skewness, and minimizing variance. People interested in considering skewness prefer a portfolio with a higher probability of large payoffs, when mean and variance remain the same. The importance of higher order moments in portfolio selection has been suggested by Samuelson (1958) as early as late 1950's. However, because of the difficulties in estimating the third order moment for a large (over a few hundred) number of securities, and in solving the non-concave function by using standard computational methodologies, quantitative treatments of the third order moment have been neglected for a long time. Since high performance computers are becoming cheaper, considering skewness in portfolio analysis is now expected to become feasible in the near future. The main question here is whether introducing skewness would significantly improve the quality of the chosen portfolios. Starting in the 1990s, several quantitative analysis have been carried out to study the optimal portfolio, taking skewness into consideration. Konno and Suzuki (1995) applied piecewise linear approximation to obtain solutions in this model. But this has only resulted in approximating the solutions. Chunhachinda, et al. (1997), have found that the returns of major stock markets all over the world are not normally distributed. They have shown that taking skewness into consideration in portfolio selection will result in a major change in the optimal portfolio.

Most of the existing portfolio selection models are based on the probability theory. One can refer to Wang and Xia (2002) for details of portfolio modelling. Though often applied to deal with uncertainty, the probabilistic approaches only partly capture the reality. Some other techniques have also been applied to handle the uncertainty of the financial markets; for instance, the fuzzy set theory [Zadeh (1965)]. In reality, although many events are characterized as fuzzy by probabilistic approaches, they are not random events. The fuzzy set theory has been widely used to solve many practical problems, including financial risk management. By using fuzzy approaches, quantitative analysis, qualitative analysis, experts' knowledge and investors' subjective opinions can be better integrated into a portfolio selection model. Recently, a few authors, such as Ramaswamy (1998), Tanaka and Guo (1999) and Inuiguchi and Ramik (2000), studied fuzzy portfolio selection.

This chapter surveys the main progress in fuzzy portfolio selection. In the next section, we introduce portfolio selection models based on the fuzzy decision theory. Portfolio selection approaches using possibilistic programming and interval programming are reviewed in Sections 3 and 4, respectively.

1.2 Portfolio Selection Based on the Fuzzy Decision Theory

The fuzzy decision theory was formulated by Bellman and Zadeh (1970). Let X denote a set of alternatives to a fuzzy decision-making problem. If there are m fuzzy goals $\tilde{G}_i$ $(i = 1, \cdots, m)$ and n fuzzy constraints $\tilde{C}_j$ $(j = 1, \cdots, n)$, then the fuzzy decision is defined by the following fuzzy set of X

$$\tilde{D} = \tilde{G}_1 \bigcap \cdots \bigcap \tilde{G}_m \bigcap \tilde{C}_1 \bigcap \cdots \bigcap \tilde{C}_n$$

with the membership function

$$\mu_{\tilde{D}}(x) = \min\{\mu_{\tilde{G}_1}(x) \bigcap \cdots \bigcap \mu_{\tilde{G}_m}(x) \bigcap \mu_{\tilde{C}_1}(x) \bigcap \cdots \bigcap \mu_{\tilde{C}_n}(x)\}.$$

Furthermore, the optimal decision is defined by the following non-fuzzy subset

$$D^O = \{x^* \in X | x^* \in \operatorname{argmax} \mu_{\tilde{D}}(x)\}.$$

One can refer to Bellman and Zadeh (1970) and Zimmermann (1985) for a detailed discussion on the fuzzy decision theory.

Ramaswamy (1998) presented a portfolio selection method using the fuzzy decision theory. The main idea is as follows. An investor can construct a portfolio based on m potential market scenarios from an investment universe of n assets with $x_i^{\min}$ and $x_i^{\max}$ being the minimum and the maximum weight of the ith asset, respectively. Let R_{ik} denote the return of the ith asset for the kth market scenario and let $R_k(x) = \sum_{i=1}^{n} R_{ik}x_i$ denote the portfolio return for the kth scenario, at the end of the investment period. For each scenario, the investor may have a target range for the expected return, over the investment period. Denoting $R_k^{\min}$ and $R_k^{\max}$ as the minimum and the maximum expected returns, respectively, for the kth market scenario, and characterizing the degree of the investor's satisfaction with portfolio x for the kth scenario as the following linear membership function

$$\mu_k(R_k(x)) = \begin{cases} 0, & \text{if } R_k(x) \leq R_k^{\min} \\ \frac{R_k(x)-R_k^{\min}}{R_k^{\max}-R_k^{\min}}, & \text{if } R_k^{\min} < R_k(x) \leq R_k^{\max} \\ 1, & \text{if } R_k(x) > R_k^{\max} \end{cases}$$

Ramaswamy (1998), based on the fuzzy decision theory, formulated the following portfolio selection model:

$$\max_{x} \quad \mu_1(R_1(x)) \bigcap \cdots \bigcap \mu_m(R_m(x))$$
$$\text{subject to}$$
$$\sum_{i=1}^{n} x_i = 1$$
$$x_i^{\min} \leq x_i \leq x_i^{\max}, \quad i = 1, \cdots n,$$

which is equivalent to the following linear programming problem:

$$\max_{x,\lambda} \quad \lambda$$
$$\text{subject to}$$
$$\mu_k(R_k(x)) \geq \lambda, \quad k = 1, \cdots, m$$
$$\sum_{i=1}^{n} x_i = 1$$
$$x_i^{\min} \leq x_i \leq x_i^{\max}, \quad i = 1, \cdots n.$$

Ramaswamy (1998) gave a numerical example in which the investor is only allowed to hold government bonds and plain vanilla options, and only two scenarios are assumed: "bullish" and "bearish".

A similar approach for portfolio selection, using the fuzzy decision theory, was proposed by León et al. (2000, 2002). Using the fuzzy decision principle, Östermark (1996) proposed a dynamic portfolio management model by fuzzifying the objectives and the constraints.

Watada (2001) presented another type of portfolio selection model, using the fuzzy decision principle. The model is directly related to the mean-variance model, where the goal rate (or the satisfaction degree) for an expected return, and the corresponding risk, are described by logistic membership functions. Characterizing the goal rate for an expected return as the membership function

$$\mu_E(r^T x) = \frac{1}{1 + \exp(-\beta_E(r^T x - E_M))}$$

and the goal rate for risk as the membership function

$$\mu_V(x^T \Sigma x) = \frac{1}{1 + \exp(\beta_V(x^T \Sigma x - V_M))},$$

Watada formulated the following portfolio selection model:

$$\max_{x,\lambda} \quad \lambda$$

subject to

$$\lambda + \exp(-\beta_E(r^T x - E_M))\lambda \le 1$$
$$\lambda + \exp(\beta_V(x^T \Sigma x - V_M))\lambda \le 1$$
$$\sum_{i=1}^{n} x_i = 1$$
$$x_i \ge 0, \quad i = 1, \cdots n.$$

where β_E and β_V are two positive parameters to determine the shapes of the membership functions, and E_M and V_M are the mid points, whose membership values are 0.5. The larger the values of β_E and β_V are, the lower will be the fuzziness.

Watada's model can be further simplified; the portfolio strategy generated by the model is also a mean-variance efficient strategy.

1.3 Portfolio Selection Based on Possibilistic Programming

In the possibility theory, proposed by Zadeh (1978) and advanced by Dubois and Prade (1988), fuzzy variables are associated with possibility distributions, similar to the way random variables are associated with probability distributions in the probability theory. According to the possibility theory, a possibility variable is represented as a convex and normal fuzzy set. The possibility distribution function of a fuzzy variable is usually defined by the membership function of the corresponding fuzzy set.

Now we introduce several notions for consequent discussion. Denote $\pi_A(x)$ as the possibility distribution function of fuzzy variable a and denote $\mu_B(x)$ as the membership function of fuzzy set B. Possibility and necessity measures of the event a being in fuzzy set B are defined as:

$$\Pi_A(B) = \sup_x \min\{\pi_A(x), \mu_B(x)\},$$

and

$$N_A(B) = \inf_x \max\{1 - \pi_A(x), \mu_B(x)\},$$

where $\Pi_A(B)$ measures the extent to which it is possible that the fuzzy variable a is in fuzzy set B, and $N_A(B)$ measures the extent to which it is certain that the fuzzy variable a is in fuzzy set B. Clearly, while B is a crisp set, $\Pi_A(B)$ and $N_A(B)$ reduce to

$$\Pi_A(B) = \sup_{x \in B} \pi_A(x)$$

and

$$N_A(B) = \inf_{x \notin B} [1 - \pi_A(x)].$$

Let g be a real number, let Pos$(a \geq g)$ and Nec$(a \geq g)$ denote the possibility and the necessity of event $a \geq g$; then we can easily obtain the Possibility and Necessity Measures

$$\text{Pos}(a \geq g) = \Pi_A([g, +\infty))$$
$$= \sup_{x}\{\pi_A(x)|x \geq g\}$$

and

$$\text{Nec}(a \geq g) = N_A([g, +\infty))$$
$$= 1 - \sup_{x}\{\pi_A(x)|x < g\}.$$

Pos$(a \leq g)$ and Nec$(a \leq g)$ can be similarly deduced.

1.3.1 The Center-Spread Model

Suppose that the uncertain return vector R of n assets in a financial market is a fuzzy vector denoted by A. Assume that A follows a possibility distribution characterized by the following exponential distribution function

$$\pi_A(R) = \exp\{-(R - c)^T D_A^{-1}(R - c)\} = (c, D_A)_e,$$

where c is a center vector and D_A is a symmetric positive-definite matrix. The possibility return of a portfolio $x = (x_1, \cdots, x_n)^T$ can be written as $z = R^T x$. The possibility distribution function of z, denoted by $\pi_Z(z)$, can be defined by the extension principle as

$$\pi_Z(z) = \max_{R}\{\pi_A(R)|z = R^T x\}.$$

Solving the above optimization problem, we can easily get

$$\pi_Z(z) = \exp\{-(z - c^T x)^2/(x^T D_A x)\},$$

where $c^T x$ is the center value and $x^T D_A x$ is the spread of the possibility return z.

Obviously, the value of the possibility measure of $c^T x$ is 1, i.e., $c^T x$ is the most possible portfolio return. $x^T D_A x$ is a measure of uncertainty (risk) of the possibility return. The larger the value of $x^T D_A x$ is, higher will be the uncertainty of the portfolio's return. The center value and spread of a possibility return are analogous to the mean and variance of a probability return in the Markowitz model.

Following Markowitz's mean-variance methodology, Tanaka and Guo (1999) formulated the following portfolio selection model:

$$\min_{x} \quad x^T D_A x$$

subject to
$$c^T x \geq r_c$$
$$\sum_{i=1}^{n} x_i = 1$$
$$x_i \geq 0, \quad i = 1, \cdots n.$$

Another form of the possibility portfolio selection model that maximizes the center return subject to a given spread level, can also be easily formulated as the Markowitz's model.

According to the mean-variance efficiency, Tanaka and Guo (1999) defined the center-spread efficiency. Varying r_c and solving the above problem, we can obtain the efficient frontier in the center-spread plane which is very similar to the mean-variance efficient frontier.

In the original model of Tanaka and Guo (1999), $c^T x \geq r_c$ is replaced by $c^T x = r_c$. But, we believe that the above model should be more suitable since the original one may generate an inefficient portfolio if the specified value r_c is too small.

In the mean-variance model, the mean vector and the covariance matrix can be easily estimated by the historical data, statistically. However, the parametric matrix D_A of the possibility distribution is much more difficult to estimate. For the given data (x_i, h_i) $(i = 1, \cdots, m)$, where $x_i = (x_{i1}, \cdots, x_{in})^T$ is a vector of returns of n assets at the ith period and h_i is an associated possibility grade given by some experts to reflect the degree of between the future state of the financial market and the ith sample, Tanaka and Guo (1999) and Tanaka et al. (2000) proposed lower and upper approximations of the exponential possibility distribution.

Suppose that c^* is the estimation of c (it can be easily estimated). The estimation of D_A by the lower approximation (the upper approximation is similar) is determined by solving the following optimization problem:

$$\min_{D_A} \quad \sum_{i=1}^{m} y_i^T D_A^{-1} y_i$$

subject to
$$y_i^T D_A^{-1} y_i \geq -\ln(h_i), \quad i = 1, \cdots, m$$
$$D_A \succ 0$$

where $y_i = x_i - c^*$, and $D_A \succ 0$ means that D_A is positive definite.

It is difficult to solve the above optimization problem. Tanaka and Guo (1999) and Tanaka et al. (2000) proposed a method based on orthogonal conditions and a rotation method, using the principal component analysis

(PCA) to simplify and relax the optimization problem, and then solve the relaxed problem to get an approximate solution. However, the rapidly developing semi-definite programming seems to be a powerful method that can be used to solve the above problem (one can refer to Vandenberghe and Boyd (1996) and Sturm (1999) for details).

One may ask how we can determine the possibility grade h_i properly. Analytic Hierarchy Process (AHP) [Saaty (1980)] may be a good alternative, since it combines the qualitative and quantitative analysis efficiently.

Carlsson, Fullér and Majlender (2002) assume that (i) each investor can assign a welfare, or utility, score to competing investment portfolios based on the expected return and risk of the portfolios; and (ii) the rates of return on securities are modelled by possibility distributions rather than probability distributions. They presented an algorithm of complexity $O(n^3)$ for finding an exact optimal solution (in the sense of utility scores) to the n-asset portfolio selection problem under possibility distributions.

1.3.2 Models Using the Necessity Measure

In the same year when Markowitz's mean-variance formulation appeared, Roy (1952) published his safety-first model for portfolio selection, which minimizes the probability of a random event causing the return of a portfolio to drop below a predetermined level. Although Roy's work did not attract as much attention as that of Markowitz's work, it is not only a pioneering work in stochastic programming, but also something like the Value-at-Risk (VaR) methodology that is widely used in modern financial risk management [Philippe (1997)]. Stochastic optimization approaches such as Roy (1952) and Kataoka (1963) can be directly used in possibility portfolio selection.

In contrast with the criterion of Roy (1952), one can maximize the necessity of the fuzzy event when the portfolio return is greater than or equal to a predetermined value. With this idea, we can formulate a portfolio selection model as

$$
\max_{x} \quad \text{Nec}(\sum_{i=1}^{n} R_i x_i \geq \gamma)
$$
$$
\text{subject to } \sum_{i=1}^{n} x_i = 1
$$
$$
x_i \geq 0, \quad i = 1, \cdots, n,
$$

where $\gamma \in (0, 1]$ is a predetermined value.

Similar to the approach of Kataoka (1963), one can formulate a portfolio selection model which maximizes γ such that the necessity of the event that the portfolio return is greater than or equal to γ is, at least, λ. Mathematically, that is

$$\max_{x,\gamma} \quad \gamma$$

$$\text{subject to Nec}(\sum_{i=1}^{n} R_i x_i \geq \gamma) \geq \lambda$$

$$\sum_{i=1}^{n} x_i = 1$$

$$x_i \geq 0, \quad i = 1, \cdots, n,$$

where $\lambda \in (0, 1]$ is a predetermined value.

Different from Tanaka and Guo (1999) and Tanaka et al. (2000), Inuiguchi and Ramik (2000) assumed that the fuzzy variables of returns of n assets are mutually independent, and follow the exponential distributions as

$$\pi_{r_i}(R_i) = \exp(-(R_i - c_i)^2/w_i^2), \quad i = 1, \cdots, n,$$

where c_i and w_i are the center and spread of the possibility distribution of return of asset i (the spread can be also defined by w_i^2 in accordance with the definition of Tanaka et al.). Under this assumption, the above two models can be easily transformed into two linear programming models [Inuiguchi and Ramik (2000)]. Generally, these similar models, based on the criteria of Roy (1952) and Kataoka (1963), using probability theory, can not be transformed into any linear programming model, even when the random returns are independently distributed.

Unfortunately, the above two models usually generate only a concentrated investment strategy, i.e., they allow only one asset as the optimal portfolio, since they are actually two linear optimization models with one constraint, except the non-negative constraints on the variables x_i ($i = 1, \cdots, n$). Therefore, these two models can not meet the needs of an investor who wants to diversify his/her investment risk with a distributive investment strategy.

However, Inuiguchi and Ramik (2000) also presented a spread minimization model, which is similar to the center-spread model proposed by Tanaka and Guo (1999) and Tanaka et al. (2000), except the assumption independence of the possibility distribution. It was shown that the model is equivalent to the mean-variance model where the coefficients of correlation between the random return variables are all equal to 1, which can be interpreted as the correlations being unknown and the portfolio risk, as measured by the variance, being an estimate of the worst case scenario only. In the model of Inuiguchi and Ramik (2000), the spread of a portfolio return is expressed as a linear term, under the assumption of independence of the possibility distribution, whereas it is a quadratic term in the model of Tanaka and Guo (1999) and Tanaka et al. (2000), without the independence assumption. The model suggests, at the most, two assets for an investor to hold, which is still a concentrated investment strategy.

Inuiguchi and Tanino (2000) discussed the cases where the uncertain returns independently follow more general possibility distributions. The models can also be transformed into linear programming models with one or two constraints, except the non-negativity restriction on the variables x_i ($i = 1, \cdots, n$). Thus, the issue of concentrated investment remains.

To model the issue of portfolio choice under a pessimistic environment, or the option of a conservative portfolio manager, Inuiguchi and Ramik (2000) and Inuiguchi and Tanino (2000) presented a minimax regret model. The optimal investment strategy generated by the model is always a dispersive one.

When an investor is informed about the determined returns $\bar{R}$ on his/her investment x at the end of the planning investment horizon, he/she will have a regret $\theta(x; \bar{R})$, which can be quantified as

$$\theta(x; \bar{R}) = \max_y \{\bar{R}^T y - \bar{R}^T x | \sum_{i=1}^n y_i = 1, y_i \geq 0, i = 1, \cdots, n\}.$$

It is the difference between return on the optimal portfolio and the real one. However, at the decision-making stage, an investor cannot know exactly the real return $\bar{R}$, revealed in the future.

Denote the possibility distribution function of the uncertain return vector as $\pi_r(R)$. By the extension principle, the possibility distribution of regret with respect to x, denoted by $\pi_{\Theta(x)}(\theta)$, can be defined by

$$\pi_{\Theta(x)}(\theta) = \max_R \{\pi_r(R) | \theta = \theta(x; R)\}.$$

Minimizing γ such that the necessity of the regret being less than or equal to γ is at least some predetermined value, we can formulate a portfolio selection model as

$$\min_{x,\gamma} \quad \gamma$$
$$\text{subject to } \text{Nec}(\theta \leq \gamma) \geq \lambda$$
$$\sum_{i=1}^n x_i = 1$$
$$x_i \geq 0, \quad i = 1, \cdots, n,$$

where $\lambda \in (0, 1]$ is a predetermined value.

The minimax regret model can be transformed into a linear programming model, under an independent distribution assumption [Inuiguchi and Ramik (2000) and Inuiguchi and Tanino (2000)].

As the necessity measure used in the above models, the possibility measure can also be similarly used to model portfolio selection problems. However, it is clear, a fuzzy event may fail, even though its possibility achieves 1, and holds, even though its necessity is 0. The credibility measures, defined by the average of possibility and necessity measures, might deserve to be used in portfolio modeling. A fuzzy event must hold if its credibility achieves 1, and fail if its credibility is 0. One can refer to Liu (2002) for details.

1.4 Portfolio Selection Based on Interval Programming

Interval $[a, b]$ is a special fuzzy number whose membership function takes value 1 over $[a, b]$, and 0 anywhere else, which is named as the interval number in terms of fuzzy mathematics. In this section, we discuss portfolio selection using interval programming. Here, we call a programming problem involving intervals as an interval programming problem. Some interval programming problems are a special type of possibilistic programming problems.

Compared with probability variables or fuzzy variables with complex distribution functions, intervals can be treated more easily. Thus it is a good alternative to formulate an optimization problem with an interval model under some uncertain environments.

Lai *et al.* (2002) extended the Markowitz model to an interval programming model by quantifying the expected return and the covariance as intervals [also see Zeng *et al.* (2002)]. Denote the inexact expected return and covariance as the following intervals:

$$\widetilde{r}_i = [r_i - \delta_{il}, r_i + \delta_{ir}],$$
$$\widetilde{\sigma}_{ij} = [\sigma_{ij} - \delta_{ijl}, \sigma_{ij} + \delta_{ijr}].$$

By interval computing, the interval portfolio return and portfolio variance are as follows:

$$\widetilde{r}(x) = [r(x) - \delta_{RL}(x), r(x) + \delta_{RR}(x)],$$
$$\widetilde{\sigma^2}(x) = [\sigma^2(x) - \delta_{VL}(x), \sigma^2(x) + \delta_{VR}(x)],$$

where

$$r(x) - \delta_{RL}(x) = \sum_{i=1}^{n} (r_i - \delta_{il})x_i,$$

$$r(x) + \delta_{RR}(x) = \sum_{i=1}^{n} (r_i + \delta_{ir})x_i,$$

$$\sigma^2(x) - \delta_{VL}(x) = \sum_{i=1}^{n}\sum_{j=1}^{n} (\sigma_{ij} - \delta_{ijl})x_i x_j,$$

$$\sigma^2(x) + \delta_{VR}(x) = \sum_{i=1}^{n}\sum_{j=1}^{n} (\sigma_{ij} + \delta_{ijr})x_i x_j.$$

According to the Markowitz model, to strike a trade-off between the mean and variance, an investor can construct his/her portfolio by solving the following optimization problem:

$$\min_{x} \quad \widetilde{\sigma^2}(x) - \alpha\widetilde{r}(x)$$
$$\text{subject to } \sum_{i=1}^{n} x_i = 1$$
$$x_i \geq 0, \quad i = 1, \cdots, n,$$

where the objective function is an interval function $[F_l(x), F_r(x)]$ satisfying

$$F_l(x) = \sigma^2(x) - \delta_{VL}(x) - \alpha(r(x) + \delta_{RR}(x)),$$
$$F_r(x) = \sigma^2(x) + \delta_{VR}(x) - \alpha(r(x) - \delta_{RL}(x)),$$

and the minimization can be interpreted as an optimization problem defined on the basis of some order of relations between intervals. One can refer to Lai *et al.* (2002) and its references for details on interval programming.

The following are two simple models. The first is as follows:

$$\min_{x} \quad F_l(x)$$
$$\text{subject to } \sum_{i=1}^{n} x_i = 1$$
$$x_i \geq 0, \quad i = 1, \cdots, n.$$

From the objective function $F_l(x)$, we can observe that the investor estimates the return and risk optimistically. The second model is

$$\min_{x} \quad F_r(x)$$
$$\text{subject to } \sum_{i=1}^{n} x_i = 1$$
$$x_i \geq 0, i = 1, \cdots, n.$$

In contrast to the first model, here, the investor estimates return and risk pessimistically.

If the investor estimates the future financial market neither too optimistically nor too pessimistically, he/she can combine the two models to construct his/her portfolio. In the above interval programming model, the probability theory and the fuzzy set theory are integrated.

Given a fuzzy number $\tilde{N}$, the expected interval defined by Heilpern (1992) is

$$EI(\tilde{N}) = \left[\int_0^1 n_\alpha^L d\alpha, \int_0^1 n_\alpha^R d\alpha \right],$$

where n_α^L and n_α^R are the left and right ends, of α-cut of the fuzzy number. Parra *et al.* (2001) developed a goal programming (GP) model for portfolio selection, based on the expected intervals of fuzzy numbers that define the objectives and target values. Given fuzzy objectives $\tilde{Z}_k$ ($k = 1, \cdots, p$) and fuzzy target values $\tilde{g}_k$ ($k = 1, \cdots, p$), Parra *et al.*'s GP approach is to find a solution such that

$$EI(\tilde{Z}_k) \approx EI(\tilde{g}_k), \quad k = 1, \cdots, p,$$

where "$\approx$" means "approximately equal".

By introducing some concepts and operations of intervals, such as "distance" and "difference", Parra *et al.* transformed their model into the following one (the model takes into account three criteria: return, risk and liquidity):

$$\min_{x,n^L,n^R,p^L,p^R,v} \sum_{k=1}^{3} v_k$$

$$\begin{aligned}
\text{subject to} \quad & p_k^L \le v_k, p_k^R \le v_k, \quad k = 1,2,3 \\
& r^T x + n_1^L - p_1^L = EI(\widetilde{r})^L \\
& r^T x + n_1^R - p_1^R = EI(\widetilde{r})^R \\
& x^T \Sigma x + n_2^L - p_2^L = EI(\widetilde{\sigma})^L \\
& x^T \Sigma x + n_2^R - p_2^R = EI(\widetilde{\sigma})^R \\
& \sum_{i=1}^{n} EI(\widetilde{l}(g(i)))^L x_i + n_3^L - p_3^L = EI(\widetilde{l})^L \\
& \sum_{i=1}^{n} EI(\widetilde{l}(g(i)))^R x_i + n_3^R - p_3^R = EI(\widetilde{l})^R \\
& \sum_{i=1}^{n} x_i = 1 \\
& x \in \mathcal{F} \\
& n_k^L - p_k^L \le n_k^R - p_k^R, n_k^L, p_k^L, n_k^R, p_k^R \ge 0, \quad k = 1,2,3 \\
& x_i \ge 0, \quad i = 1, \cdots, n
\end{aligned}$$

where $\mathcal{F}$ represents a special requirement of x, $g(i)$ denotes the asset group that asset i belongs to, and $l(g(i))$ denotes the liquidity measure of an asset that belongs to group $g(i)$.

An application of the GP model to Spanish mutual funds is given in Parra et al. (2001). It is obvious that the above GP model is also an integrated model, which combines probability and fuzzy approaches.

The objectives of return and risk are not treated as fuzzy in the above GP model. It might be a good idea to combine the Lai et al.'s model and Parra et al.'s model [Lai et al. (2001a) and Parra et al. (2001)], which allows the objectives to be fuzzy.

Konno and Yamazika (1991), Speranza (1993) and Mansini and Speranza (1999) investigated linear programming models for portfolio selection by using absolute or semi-absolute deviation as the measure of risk. Taking the uncertain returns of assets as intervals, Lai et al. (2002) gave a linear interval programming model for portfolio selection. The model takes the maximum semi-absolute deviation of the samples of return as the measure of risk. Using the average semi-absolute deviation measure of risk, Fang et al. (2001) proposed another linear interval programming model. Based on some order of relationships between intervals, their model can be transformed into a traditional linear programming problem. Applications to portfolios of stocks in China stock markets were given in Lai et al. (2002) and Fang et al. (2001, 2005, 2006).

Part II

Portfolio Selection Models Based on Fuzzy Decision Making

2

Fuzzy Decision Making and Maximization Decision Making

Due to incomplete knowledge and information, it is not enough to use precise mathematics to model a complex system. In order to represent the vagueness in everyday life, Zadeh introduced the concept of fuzzy sets in 1965. Based on this concept, Bellman and Zadeh presented the fuzzy decision theory. They defined decision-making in a fuzzy environment with a decision set which unifies a fuzzy objective and a fuzzy constraint.

Suppose that fuzzy sets are defined on a set of alternatives, X. Let the fuzzy set for the fuzzy objective be identified as G, the fuzzy set for the constraints as C, and let decision set D be a unifying factor between G and C. Then, decision set D can be defined as an intersection set between fuzzy objective G and fuzzy constraint C, $D = G \cap C$. Decision set D is a fuzzy set, named a fuzzy decision. The corresponding membership function of the decision set D is given by:

$$\mu_D(x) = \min(\mu_G(x), \mu_C(x)), \forall x \in X. \tag{2.1}$$

More generally, if there are m fuzzy goals G_i $(i = 1, \cdots, m)$ and n fuzzy constraints C_j $(j = 1, \cdots, n)$, then the fuzzy decision is defined by the following fuzzy set

$$D = \{G_1 \cap G_2 \cap \cdots \cap G_m\} \cap \{C_1 \cap C_2 \cap \cdots \cap C_n\}. \tag{2.2}$$

Its membership function is characterized as in the following.

$$\mu_D(x) = \min(\mu_{G_1}(x), \cdots, \mu_{G_m}(x), \mu_{C_1}(x), \cdots, \mu_{C_n}(x)), \forall x \in X. \tag{2.3}$$

The above concepts illustrate that there is no difference between the objective and the constraint in a fuzzy environment.

Bellman and Zadeh proposed a maximization decision. The maximization decision is defined by the following non-fuzzy subset.

$$D^* = \{x^* \in X | x^* = \operatorname{argmax}\{\mu_D(x)\} = \operatorname{argmax}\{\min(\mu_G(x), \mu_C(x))\}\}. \tag{2.4}$$

Where m fuzzy objectives and n fuzzy constraints are given, the maximization decision can be denoted as follows:

$$D^* = \{x^* \in X | x^* = \text{argmax}\{\mu_D(x)\}$$
$$= \text{argmax}\{\min(\mu_{G_1}(x), \cdots, \mu_{G_m}(x), \mu_{C_1}(x), \cdots, \mu_{C_n}(x))\}\}.$$

The maximization decision can be considered to be an optimal decision.

3

Portfolio Selection Model with Fuzzy Liquidity Constraints

3.1 Introduction

In 1952, Markowitz published his pioneering work which laid the foundation of modern portfolio analysis. Markowitz's model has served as a basis for development of the modern financial theory over the past five decades. However, contrary to its theoretical reputation, it is not used extensively to construct large-scale portfolios. One of the most important reasons is the computational difficulty associated with solving a large-scale quadratic programming problem with a dense covariance matrix. Konno and Yamazaki (1991) used the absolute deviation risk function, to replace the risk function in Markowitz's model, and thus formulated a mean absolute deviation portfolio optimization model. It turns out that the mean absolute deviation model retains the useful properties of Markowitz's model and removes most of the principal computational difficulties in solving Markowitz's model. Simaan (1997) provided a thorough comparison of the mean variance model and the mean absolute deviation model. Furthermore, Speranza (1993) used semi-absolute deviation to measure the risk and formulated a portfolio selection model.

Transaction cost is one of the main sources of concern to portfolio managers. Arnott and Wagner (1990) found that ignoring transaction costs would result in an inefficient portfolio. Yoshimoto's empirical analysis (1996) also drew the same conclusion. Mao (1970), Jacob (1974), Brennan (1993), Levy (1978), Patel and Subhmanyam (1982), Morton and Pliska (1995) and Mansini and Speranza (1999) studied portfolio optimization with fixed transaction costs; Pogue (1986), Chen, Jen and Zionts (1971) and Yoshimoto (1996) *et al.* studied portfolio optimization with variable transaction costs; Dumas and Luciano (1991), Mulvey and Vladimirou (1992) and Dantzig and Infanger (1993) incorporated the transaction costs into the multi-period portfolio selection model. Very recently, Li, Wang and Deng (2000) gave a linear programming algorithm to solve a general mean variance model for portfolio selection with transaction costs. Konno and Wijayanayake (2001) studied portfolio optimization with transaction costs which can be expressed approximately as a D.C. function.

Usually, expected return and risk are the two fundamental factors which investors consider. In some cases, investors may consider other factors such as liquidity. Liquidity has been measured as the degree of probability of having the option of conversion of an investment into cash without any significant loss in value. Parra, Terol and Uría (2001) took into account three criteria (return, risk and liquidity) and used a fuzzy goal programming approach to solve the portfolio selection problem.

In this chapter, we will propose portfolio selection models with fuzzy liquidity constraints. At first, we construct a risk function - Minimax semi-absolute deviation risk function. Then we will formulate two optimization models for a portfolio selection problem with fuzzy liquidity constraints, based on the new risk function and semi-absolute deviation risk function, respectively.

3.2 Minimax Semi-absolute Deviation Risk Function

Speranza and Mansini used the mean semi-absolute deviation to measure the risk and formulated a mean semi-absolute deviation portfolio selection model. In the following we will propose a new risk function based on the Minimax rule.

Let $x^+ = (x_1^+, x_2^+, \cdots, x_n^+)$ and $x^- = (x_1^-, x_2^-, \cdots, x_n^-)$, where x_i^+ is the amount of the asset $i, i = 1, 2, \cdots, n$ bought by the investor, x_i^- is the amount of the asset $i, i = 1, 2, \cdots, n$ for which the asset has been sold by the investor.

Let portfolio $x = (x_1, x_2, \cdots, x_n)$, where x_i is the proportion of the security owned by the investor, such that

$$\sum_{i=1}^{n} x_i = 1.$$

Assume we have observed historical data of n securities, over T horizons. Let r_{it} be the historical return of security i $(i = 1, 2, \cdots, n)$ at t $(t = 1, 2, \cdots, T)$, r_i is the expected return of security i $(i = 1, 2, \cdots, n)$, i.e.,

$$r_i = \frac{1}{T} \sum_{t=1}^{T} r_{it}, \quad i = 1, 2, \cdots, n,$$

Then the expected return of portfolio $x = (x_1, x_2, \cdots, x_n)$ is

$$r(x) = \sum_{i=1}^{n} r_i x_i.$$

The semi-absolute deviation of return on portfolio $x = (x_1, x_2, \cdots, x_n)$ below the expected return over the past period t, $t = 1, 2, \cdots, T$ can be represented as

$$w_t(x) = \left| \min \left\{ 0, \sum_{i=1}^{n}(r_{it} - r_i)x_i \right\} \right| = \frac{\left| \sum_{i=1}^{n}(r_{it} - r_i)x_i \right| - \sum_{i=1}^{n}(r_{it} - r_i)x_i}{2}.$$

So the expected semi-absolute deviation of the return on portfolio $x = (x_1, x_2, \cdots, x_n)$, below the expected return, can be represented as

$$Smad(x) = \frac{1}{T}\sum_{t=1}^{T} w_t(x).$$

$Smad(x)$ is the semi-absolute deviation risk function proposed by Speranza.

Based on the Minimax rule, we define the maximum of T semi-absolute deviations of return on portfolio $x = (x_1, x_2, \cdots, x_n)$, below the expected return, over all the past period t, $t = 1, 2, \cdots, T$ as risk, i.e.,

$$Minimaxmad(x) = \max_t \{w_t(x), \quad t = 1, 2, \cdots, T\}$$
$$= \max_t \left\{ \left| \min\{0, \sum_{i=1}^{n}(r_{it} - r_i)x_i\} \right|, t = 1, 2, \cdots, T \right\}.$$

It is obvious that the minimax semi-absolute deviation risk function is more pessimistic than the semi-absolute deviation risk function.

3.3 Fuzzy Liquidity of Securities

Liquidity has been measured as the degree of probability of having an option to convert an investment into cash without any significant loss in value. The turnover rate of a security is the proportion of turnover volume to tradable volume of the security, and is a factor which may reflect the security's liquidity. Generally, investors prefer greater liquidity, especially since in a bull market for securities, returns on securities with high liquidity tend to increase with time. Here, we use the turnover rates of securities to measure their liquidity. It is known that turnover rates of securities in the future cannot be accurately predicted in a securities market. The possibility theory has been proposed by Zadeh and advanced by Dubois and Prade where fuzzy variables are associated with the possibility distribution. In this study, we assume that the turnover rates of securities are modeled by possibility distributions rather than by probability distributions. That is, the turnover rates of the securities will be represented by fuzzy numbers. In many cases, it might be easier to estimate the possibility distributions of turnover rates of securities, rather than the corresponding probability distributions. In this study, we regard trapezoidal possibility distribution as the possibility distribution of the turnover rates of the securities.

A fuzzy number A is called trapezoidal with tolerance interval $[a, b]$, left width α and right width β, if its membership function takes the following form:

$$A(t) = \begin{cases} 1 - \frac{a-t}{\alpha} & \text{if } a - \alpha \le t \le a, \\ 1 & \text{if } a \le t \le b, \\ 1 - \frac{t-b}{\beta} & \text{if } a \le t \le b + \beta, \\ 0 & \text{otherwise} \end{cases} \tag{3.1}$$

and we denote $A = (a, b, \alpha, \beta)$. It can easily be shown that

$$[A]^\gamma = [a - (1 - \gamma)\alpha, b + (1 - \gamma)\beta], \forall \gamma \in [0, 1], \tag{3.2}$$

where $[A]^\gamma$ denotes the γ-level set of A.

Let $[A]^\gamma = [a_1(\gamma), a_2(\gamma)]$ and $[B]^\gamma = [b_1(\gamma), b_2(\gamma)]$ be fuzzy numbers and let $k \in R$ be a real number. Using the extension principle we can verify the following rules for addition and scalar multiplication of fuzzy numbers:

$$[A + B]^\gamma = [a_1(\gamma) + b_1(\gamma), a_2(\gamma) + b_2(\gamma)], \tag{3.3}$$

$$[kA]^\gamma = k[A]^\gamma. \tag{3.4}$$

Carlsson and Fullér (2001) introduced the notation of crisp possibilistic mean value and crisp possibilistic variance of continuous possibility distributions, which are consistent with the extension principle. The crisp possibilistic mean value of A is

$$E(A) = \int_0^1 \gamma(a_1(\gamma) + a_2(\gamma))d\gamma. \tag{3.5}$$

It is clear that if $A = (a, b, \alpha, \beta)$ is a trapezoidal fuzzy number, then

$$E(A) = \int_0^1 \gamma[a - (1 - \gamma)\alpha + b + (1 - \gamma)\beta]d\gamma = \frac{a + b}{2} + \frac{\beta - \alpha}{6} \tag{3.6}$$

Denote the turnover rate of security j by the trapezoidal fuzzy number $\hat{l}_j = (la_j, lb_j, \alpha_j, \beta_j)$. Then the turnover rate of portfolio $x = (x_1, x_2, \cdots, x_n)$ is $\sum_{j=1}^n \hat{l}_j x_j$.

By the definition, the crisp possibilistic mean value of the turnover rate of security j is represented as follows:

$$E(\hat{l}_j) = \int_0^1 \gamma[la_j - (1 - \gamma)\alpha_j + lb_j + (1 - \gamma)\beta_j]d\gamma = \frac{la_j + lb_j}{2} + \frac{\beta_j - \alpha_j}{6}. \tag{3.7}$$

Therefore, the crisp possibilistic mean value of the turnover rate of portfolio $x = (x_1, x_2, \cdots, x_n)$ can be represented as

$$E(\hat{l}(x)) = E(\sum_{j=1}^n \hat{l}_j x_j) = \sum_{j=1}^n (\frac{la_j + lb_j}{2} + \frac{\beta_j - \alpha_j}{6})x_j. \tag{3.8}$$

In the study, we use the crisp possibilistic mean value of the turnover rate to measure the portfolio liquidity.

3.4 Model Formulation

Suppose an investor allocates his/her wealth among n securities offering random rates of return and a risk-less asset offering a fixed rate of return. The investor starts with an existing portfolio and decides how to reallocate assets. We introduce some notations as follows.

r_i: the expected rate of return of risky asset i $(i = 1, 2, \cdots, n)$;

r_{n+1}: the rate of return of risk-free asset $n + 1$;

x_i: the proportion of the total investment devoted to risky asset i $(i = 1, 2, \cdots, n)$ or risk-free asset $n + 1$;

x_i^0: the proportion of the risky asset i $(i = 1, 2, \cdots, n)$ or risk-free asset $n + 1$ owned by the investor;

r_{it}: the historical rate of return of risky asset i $(i = 1, 2, \cdots, n)$, t $(t = 1, 2, \cdots, T)$;

k_i: the rate of transaction costs for asset i $(i = 1, 2, \cdots, n + 1)$;

u_i: the upper bound of proportion of the total investment devoted to risky asset i $(i = 1, 2, \cdots, n)$ or risk-free asset $n + 1$.

We use V shape function to express transaction costs, so the transaction costs of the asset i $(i = 1, 2, \cdots, n + 1)$ can be denoted by

$$C_i(x_i) = k_i |x_i - x_i^0|. \tag{3.9}$$

So the total transaction costs of portfolio $x = (x_1, x_2, \cdots, x_n, x_{n+1})$ can be denoted by

$$C(x) = \sum_{i=1}^{n+1} C_i(x_i) = \sum_{i=1}^{n+1} k_i |x_i - x_i^0|. \tag{3.10}$$

The expected return of portfolio $x = (x_1, x_2, \cdots, x_{n+1})$ in the future can be represented as

$$r(x) = \sum_{i=1}^{n+1} r_i x_i.$$

We can use the arithmetic mean of historical data as the expected return, i.e.,

$$r_i = \frac{1}{T} \sum_{t=1}^{T} r_{it}, \quad i = 1, 2, \cdots, n.$$

After removing the transaction costs, the net expected return of portfolio $x = (x_1, x_2, \cdots, x_{n+1})$ can be represented as

$$f(x) = \sum_{i=1}^{n+1} (r_i x_i - k_i |x_i - x_i^0|).$$

The semi-absolute deviation of the return of portfolio $x = (x_1, x_2, \cdots, x_{n+1})$ below the expected return at the past period t, $t = 1, 2, \cdots, T$ can be represented as

$$Smad(x) = \frac{1}{T} \sum_{t=1}^{T} \left| \min\{0, \sum_{i=1}^{n} (r_{it} - r_i)x_i\} \right|.$$

The Minimax semi-absolute deviation of the return of portfolio $x = (x_1, x_2, \cdots, x_{n+1})$ below the expected return at the past period t, $t = 1, 2, \cdots, T$ can be represented as

$$Minimaxmad(x) = \max_t \left\{ \left| \min\{0, \sum_{i=1}^{n} (r_{it} - r_i)x_i\} \right|, \ t = 1, 2, \cdots, T \right\}.$$

Denote the turnover rate of security j by trapezoidal fuzzy number $\hat{l}_j = (la_j, lb_j, \alpha_j, \beta_j)$. Then the turnover rate of portfolio $x = (x_1, x_2, \cdots, x_n)$ is $\sum_{j=1}^{n} \hat{l}_j x_j$.

Assume the investor wants to maximize return and minimize risk after paying transaction costs. Based on the above discussions, the portfolio selection problem is formulated as the following bi-objective programming problems.

If we use the semi-absolute deviation risk function to measure risk, then we can get the following bi-objective programming problem

(BP3-1) $\max \ f(x) = \sum_{i=1}^{n+1} (r_i x_i - k_i |x_i - x_i^0|),$

$$\min \ Smad(x) = \frac{1}{T} \sum_{t=1}^{T} \left| \min\{0, \sum_{i=1}^{n} (r_{it} - r_i)x_i\} \right|$$

s.t. $\sum_{i=1}^{n+1} x_i = 1,$

$0 \le x_i \le u_i, \quad i = 1, 2, \cdots, n+1,$

$\sum_{j=1}^{n+1} \hat{l}_j x_j \ge \hat{l}_0,$

where $\hat{l}_0$ is the tolerance level of the fuzzy turnover rate given by the investor.

If we use the Minimax semi-absolute deviation risk function to measure risk, then we can get the following bi-objective programming problem

(BP3-2) $\max \ f(x) = \sum_{i=1}^{n+1} (r_i x_i - k_i |x_i - x_i^0|),$

$$\min_x \ \max_t \left\{ \left| \min\{0, \sum_{i=1}^{n} (r_{it} - r_i)x_i\} \right|, \ t = 1, 2, \cdots, T \right\}$$

s.t. $\sum_{i=1}^{n+1} x_i = 1,$

$0 \le x_i \le u_i, \quad i = 1, 2, \cdots, n+1,$

$\sum_{j=1}^{n+1} \hat{l}_j x_j \ge \hat{l}_0,$

where $\hat{l}_0$ is the tolerance level of fuzzy turnover rate given by the investor.

A fuzzy number A is called trapezoidal with tolerance interval $[a, b]$, left width α and right width β if its membership function takes the following form:

$$A(t) = \begin{cases} 1 - \frac{a-t}{\alpha} & \text{if } a - \alpha \leq t \leq a, \\ 1 & \text{if } a \leq t \leq b, \\ 1 - \frac{t-b}{\beta} & \text{if } a \leq t \leq b + \beta, \\ 0 & \text{otherwise} \end{cases} \qquad (3.11)$$

and we denote $A = (a, b, \alpha, \beta)$. It can easily be shown that

$$[A]^\gamma = [a - (1 - \gamma)\alpha, b + (1 - \gamma)\beta], \forall \gamma \in [0, 1], \qquad (3.12)$$

where $[A]^\gamma$ denotes the γ-level set of A.

Let $[A]^\gamma = [a_1(\gamma), a_2(\gamma)]$ and $[B]^\gamma = [b_1(\gamma), b_2(\gamma)]$ be fuzzy numbers and let $k \in R$ be a real number. Using the extension principle, we can verify the following rules for addition and scalar multiplication of fuzzy numbers:

$$[A + B]^\gamma = [a_1(\gamma) + b_1(\gamma), a_2(\gamma) + b_2(\gamma)], \qquad (3.13)$$

$$[kA]^\gamma = k[A]^\gamma. \qquad (3.14)$$

Carlsson and Fullér introduced the notation of crisp possibilistic mean value and crisp possibilistic variance of continuous possibility distributions, which are consistent with the extension principle. The crisp possibilistic mean value of A is

$$E(A) = \int_0^1 \gamma(a_1(\gamma) + a_2(\gamma))d\gamma. \qquad (3.15)$$

It is clear that if $A = (a, b, \alpha, \beta)$ is a trapezoidal fuzzy number, then

$$E(A) = \int_0^1 \gamma[a - (1 - \gamma)\alpha + b + (1 - \gamma)\beta]d\gamma = \frac{a+b}{2} + \frac{\beta - \alpha}{6} \qquad (3.16)$$

Denote the turnover rate of security j by the trapezoidal fuzzy number $\hat{l}_j = (la_j, lb_j, \alpha_j, \beta_j)$. Then the turnover rate of portfolio $x = (x_1, x_2, \cdots, x_n)$ is $\sum_{j=1}^n \hat{l}_j x_j$.

By the definition, the crisp possibilistic mean value of the turnover rate of security j is represented as follows:

$$E(\hat{l}_j) = \int_0^1 \gamma[la_j - (1-\gamma)\alpha_j + lb_j + (1-\gamma)\beta_j]d\gamma = \frac{la_j + lb_j}{2} + \frac{\beta_j - \alpha_j}{6}. \qquad (3.17)$$

Therefore, the crisp possibilistic mean value of the turnover rate of portfolio $x = (x_1, x_2, \cdots, x_n)$ can be represented as

$$E(\hat{l}(x)) = E(\sum_{j=1}^{n+1} \hat{l}_j x_j) = \sum_{j=1}^{n+1} \left(\frac{la_j + lb_j}{2} + \frac{\beta_j - \alpha_j}{6} \right) x_j.$$

In the study, we use the crisp possibilistic mean value of the turnover rate to measure the portfolio liquidity.

The fuzzy inequations in (BP3-1) and (BP3-2)

$$\sum_{j=1}^{n+1} \hat{l}_j x_j \geq \hat{l}_0$$

can be transformed into

$$\sum_{j=1}^{n+1} \left(\frac{la_j + lb_j}{2} + \frac{\beta_j - \alpha_j}{6} \right) x_j \geq E(\hat{l}_0).$$

Hence, (BP3-1) and (BP3-2) can be transformed into (BP3-3) and (BP3-4)

(BP3-3) $\max f(x) = \sum_{i=1}^{n+1} (r_i x_i - k_i |x_i - x_i^0|),$

$\min \ Smad(x) = \frac{1}{T} \sum_{t=1}^{T} \left| \min\{0, \sum_{i=1}^{n} (r_{it} - r_i)x_i\} \right|$

s.t. $\sum_{i=1}^{n+1} x_i = 1,$

$0 \leq x_i \leq u_i, \quad i = 1, 2, \cdots, n+1,$

$\sum_{j=1}^{n+1} \left(\frac{la_j + lb_j}{2} + \frac{\beta_j - \alpha_j}{6} \right) x_j \geq E(\hat{l}_0);$

(BP3-4) $\max f(x) = \sum_{i=1}^{n+1} (r_i x_i - k_i |x_i - x_i^0|),$

$\min\limits_{x} \max\limits_{t} \left\{ \left| \min\{0, \sum_{i=1}^{n} (r_{it} - r_i)x_i\} \right|, \ t = 1, 2, \cdots, T \right\}$

s.t. $\sum_{i=1}^{n+1} x_i = 1,$

$0 \leq x_i \leq u_i, \quad i = 1, 2, \cdots, n+1,$

$\sum_{j=1}^{n+1} \left(\frac{la_j + lb_j}{2} + \frac{\beta_j - \alpha_j}{6} \right) x_j \geq E(\hat{l}_0),$

The above bi-objective programming problem can be solved by transforming it into a single objective programming problem.

Assuming the investor has a minimal return level on the portfolio, (BP3-3) can be transformed into (P3-1)

$$(\text{P3-1}) \quad \min \frac{1}{T} \sum_{t=1}^{T} \left| \min \left\{ 0, \sum_{i=1}^{n} (r_{it} - r_i) x_i \right\} \right|$$

$$\text{s.t.} \quad f(x) = \sum_{i=1}^{n+1} (r_i x_i - k_i |x_i - x_i^0|) \geq r_0,$$

$$\sum_{i=1}^{n+1} x_i = 1,$$

$$0 \leq x_i \leq u_i, \quad i = 1, 2, \cdots, n+1,$$

$$\sum_{j=1}^{n+1} \left(\frac{la_j + lb_j}{2} + \frac{\beta_j - \alpha_j}{6} \right) x_j \geq E(\hat{l}_0),$$

where r_0 is a given constant representing the minimal return level on the portfolio required by the investor, and $E(\hat{l}_0)$ is the level of fuzzy turnover rate.

Similarly, (BP3-4) can be transformed into (P3-2)

$$(\text{P3-2}) \quad \min_x \max_t \left\{ \left| \min\{0, \sum_{i=1}^{n} (r_{it} - r_i) x_i\} \right|, \ t = 1, 2, \cdots, T \right\}$$

$$\text{s.t.} \quad \sum_{i=1}^{n+1} (r_i x_i - k_i |x_i - x_i^0|) \geq r_0,$$

$$\sum_{i=1}^{n+1} x_i = 1,$$

$$0 \leq x_i \leq u_i, \quad i = 1, 2, \cdots, n+1,$$

$$\sum_{j=1}^{n+1} \left(\frac{la_j + lb_j}{2} + \frac{\beta_j - \alpha_j}{6} \right) x_j \geq E(\hat{l}_0),$$

where r_0 is the minimum level of return, and $E(\hat{l}_0)$ is the required level of fuzzy turnover rate.

If the investor has a tolerance level of risk of the portfolio, we can get (P3-3) and (P3-4)

(P3-3) $\max f(x) = \sum_{i=1}^{n+1}(r_i x_i - k_i |x_i - x_i^0|)$

s.t. $\frac{1}{T}\sum_{t=1}^{T}\left|\min\left\{0, \sum_{i=1}^{n}(r_{it} - r_i)x_i\right\}\right| \leq w_0,$

$\sum_{i=1}^{n+1} x_i = 1,$

$0 \leq x_i \leq u_i, \quad i = 1, 2, \cdots, n+1,$

$\sum_{j=1}^{n+1}\left(\frac{la_j + lb_j}{2} + \frac{\beta_j - \alpha_j}{6}\right)x_j \geq E(\hat{l}_0);$

(P3-4) $\max f(x) = \sum_{i=1}^{n+1}(r_i x_i - k_i |x_i - x_i^0|)$

s.t. $\max_t\left\{\left|\min\{0, \sum_{j=1}^{n}(r_{it} - r_i)x_i\}\right|, \ t = 1, 2, \cdots, T\right\} \leq w_0,$

$\sum_{i=1}^{n+1} x_i = 1,$

$0 \leq x_i \leq u_i, \quad i = 1, 2, \cdots, n+1,$

$\sum_{j=1}^{n+1}\left(\frac{la_j + lb_j}{2} + \frac{\beta_j - \alpha_j}{6}\right)x_j \geq E(\hat{l}_0).$

where w_0 is a given constant representing the tolerance level of risk, and $E(\hat{l}_0)$ is the required level of fuzzy turnover rate.

(P3-1), (P3-2) and (P3-3), (P3-4) can be used interchangeably to generate the efficient frontier of portfolios. In this sequel, we only discuss (P3-3) and (P3-4). To solve (P3-3) and (P3-4), we consider the following transformation.

At first, introducing a new variable x_{n+2}, let

$$\sum_{i=1}^{n+1} k_i |x_i - x_i^0| \leq x_{n+2},$$

we can get the following model

(P3-5) $\max f(x) = \sum\limits_{i=1}^{n+1} r_i x_i - x_{n+2}$

s.t. $\max\limits_{t} \left\{ \left| \min\{0, \sum\limits_{i=1}^{n} (r_{it} - r_i)x_i\} \right|, \ t = 1, 2, \cdots, T \right\} \leq w_0,$

$\sum\limits_{i=1}^{n+1} k_i |x_i - x_i^0| \leq x_{n+2},$

$\sum\limits_{i=1}^{n+1} x_i = 1,$

$0 \leq x_i \leq u_i, \quad i = 1, 2, \cdots, n+1,$

$\sum\limits_{j=1}^{n+1} (\frac{la_j + lb_j}{2} + \frac{\beta_j - \alpha_j}{6}) x_j \geq E(\hat{l}_0).$

Theorem 3.1   The portfolio $(x_1^*, \cdots, x_{n+1}^*)$ is an optimal solution of (P3-4), if and only if there is a variable x_{n+2}^*, such that $(x_1^*, \cdots, x_{n+1}^*, x_{n+2}^*)$ is an optimal solution of (P2-5).
Proof If $(x_1^*, \cdots, x_{n+1}^*)$ is an optimal solution of (P3-4), let

$$x_{n+2}^* = \sum_{i=1}^{n+1} k_i |x_i^* - x_i^0|,$$

then $(x_1^*, \cdots, x_{n+1}^*, x_{n+2}^*)$ is a feasible solution of (P3-5).
If $(x_1^*, \cdots, x_{n+1}^*, x_{n+2}^*)$ is not an optimal solution of (P3-5), then there is a feasible solution of (P3-5) $(x_1^{'}, \cdots, x_{n+1}^{'}, x_{n+2}^{'})$, such that

$$\sum_{i=1}^{n+1} r_i x_i^{'} - x_{n+2}^{'} > \sum_{i=1}^{n+1} r_i x_i^* - x_{n+2}^*.$$

For any feasible solutions, we have $x_{n+2} \geq \sum\limits_{i=1}^{n+1} k_i |x_i - x_i^0|$, so

$$\sum_{i=1}^{n+1} r_i x_i^{'} - k_i |x_i^{'} - x_i^0| \geq \sum_{i=1}^{n+1} r_i x_i^{'} - x_{n+2}^{'} > \sum_{i=1}^{n+1} r_i x_i^* - x_{n+2}^*$$
$$= \sum_{i=1}^{n+1} r_i x_i^* - k_i |x_i^* - x_i^0|, \tag{3.18}$$

i.e., we can find a feasible solution $(x_1^{'}, \cdots, x_{n+1}^{'})$, such that

$$\sum_{i=1}^{n+1} r_i x_i^{'} - k_i |x_i^{'} - x_i^0| > \sum_{i=1}^{n+1} r_i x_i^* - k_i |x_i^* - x_i^0|.$$

It contradicts that $(x_1^*, \cdots, x_{n+1}^*)$ is an optimal solution of (P3-4).

Contrarily, if $(x_1^*, \cdots, x_{n+1}^*, x_{n+2}^*)$ is an optimal solution of (P3-5), then it is obvious that $(x_1^*, \cdots, x_{n+1}^*)$ is a feasible solution of (P3-4). If it is not an optimal solution of (P3-4), then there is a feasible solution $(x_1^{''}, \cdots, x_{n+1}^{''})$, such that

$$\sum_{i=1}^{n+1} r_i x_i^{''} - k_i |x_i^{''} - x_i^0| > \sum_{i=1}^{n+1} r_i x_i^* - k_i |x_i^* - x_i^0|.$$

Let $x_{n+2}^{''} = \sum_{i=1}^{n+1} k_i |x_i^{''} - x_i^0|$, then we have

$$\sum_{i=1}^{n+1} r_i x_i^{''} - x_{n+2}^{''} = \sum_{i=1}^{n+1} r_i x_i^{''} - k_i |x_i^{''} - x_i^0| > \sum_{i=1}^{n+1} r_i x_i^* - k_i |x_i^* - x_i^0|$$

$$= \sum_{i=1}^{n+1} r_i x_i^* - x_{n+2}^*.$$

i.e., we can find a feasible solution $(x_1^{''}, \cdots, x_{n+2}^{''})$, such that

$$\sum_{i=1}^{n+1} r_i x_i^{''} - x_{n+2}^{''} > \sum_{i=1}^{n+1} r_i x_i^* - x_{n+2}^*$$

It contradicts that $(x_1^*, \cdots, x_{n+1}^*, x_{n+2}^*)$ is an optimal solution of (P3-5).

Since

$$\left| \min \left\{ 0, \sum_{j=1}^{n} (r_{it} - r_i) x_i \right\} \right| = \max \left\{ 0, \sum_{j=1}^{n} (r_i - r_{it}) x_i \right\}$$

$$= \frac{\left| \sum_{i=1}^{n} (r_{it} - r_i) x_i \right| - \sum_{i=1}^{n} (r_{it} - r_i) x_i}{2},$$

$$t = 1, 2, \cdots, T.$$

(P3-5) can be transformed into the following problem

(P3-6) $\max f(x) = \sum_{i=1}^{n+1} r_i x_i - x_{n+2}$

s.t. $\dfrac{\left| \sum_{j=1}^{n} (r_{it} - r_i)x_i \right| - \sum_{i=1}^{n}(r_{it} - r_i)x_i}{2} \leq w_0, \quad t = 1, 2, \cdots, T,$

$\sum_{i=1}^{n+1} k_i |x_i - x_i^0| \leq x_{n+2},$

$\sum_{i=1}^{n+1} x_i = 1,$

$0 \leq x_i \leq u_i, \quad i = 1, 2, \cdots, n+1,$

$\sum_{j=1}^{n+1} \left(\dfrac{la_j + lb_j}{2} + \dfrac{\beta_j - \alpha_j}{6} \right) x_j \geq E(\hat{l}_0).$

Let

$$d_i^+ = \frac{|x_i - x_i^0| + (x_i - x_i^0)}{2},$$

$$d_i^- = \frac{|x_i - x_i^0| - (x_i - x_i^0)}{2},$$

$$y_t^+ = \frac{\left| \sum_{i=1}^{n}(r_{it} - r_i)x_i \right| + \sum_{i=1}^{n}(r_{it} - r_i)x_i}{2},$$

$$y_t^- = \frac{\left| \sum_{i=1}^{n}(r_{it} - r_i)x_i \right| - \sum_{i=1}^{n}(r_{it} - r_i)x_i}{2}.$$

Then we have

$$d_i^+ + d_i^- = |x_i - x_i^0|,$$
$$d_i^+ - d_i^- = x_i - x_i^0,$$
$$d_i^+ d_i^- = 0,$$
$$d_i^+ \geq 0, \quad d_i^- \geq 0,$$
$$y_t^+ + y_t^- = \left| \sum_{i=1}^{n}(r_{it} - r_i)x_i \right|,$$
$$y_t^+ - y_t^- = \sum_{i=1}^{n}(r_{it} - r_i)x_i,$$
$$y_t^+ y_t^- = 0,$$
$$y_t^+ \geq 0, \quad y_t^- \geq 0.$$

Thus, (P3-6) can be rewritten as the following problem

$$\text{(P3-7)} \quad \max f(x) = \sum_{i=1}^{n+1} r_i x_i - x_{n+2}$$

$$\text{s.t.} \quad y_t^- \le w_0, \quad t = 1, 2, \cdots, T,$$

$$\sum_{i=1}^{n+1} k_i(d_i^+ + d_i^-) \le x_{n+2},$$

$$d_i^+ - d_i^- = x_i - x_i^0, \quad i = 1, 2, \cdots, n+1,$$

$$d_i^+ d_i^- = 0, \quad i = 1, 2, \cdots, n+1,$$

$$y_t^+ - y_t^- = \sum_{i=1}^{n}(r_{it} - r_i)x_i, \quad t = 1, 2, \cdots, T,$$

$$y_t^+ y_t^- = 0, \quad t = 1, 2, \cdots, T,$$

$$\sum_{i=1}^{n+1} x_i = 1,$$

$$\sum_{j=1}^{n+1}\left(\frac{la_j + lb_j}{2} + \frac{\beta_j - \alpha_j}{6}\right) x_j \ge E(\hat{l}_0),$$

$$0 \le x_i \le u_i, \quad i = 1, 2, \cdots, n+1,$$

$$d_i^+ \ge 0, \quad d_i^- \ge 0, \quad i = 1, 2, \cdots, n+1,$$

$$y_t^+ \ge 0, \quad y_t^- \ge 0, \quad t = 1, 2, \cdots, T.$$

Hence, we can obtain the following theorem:

Theorem 3.2   Portfolio $(x_1^*, \cdots, x_{n+1}^*)$ is an optimal solution of (P3-6), if and only if $d_1^{+*}, \cdots, d_{n+1}^{+*}, d_1^{-*}, \cdots, d_{n+1}^{-*}, y_1^{+*}, \cdots, y_T^{+*}, y_1^{-*}, \cdots, y_T^{-*}$, such that $(x_1^*, \cdots, x_{n+2}^*, d_1^{+*}, \cdots, d_{n+1}^{+*}, d_1^{-*}, \cdots, d_{n+1}^{-*}, y_1^{+*}, \cdots, y_T^{+*}, y_1^{-*}, \cdots, y_T^{-*})$ is an optimal solution of (P3-7).

Eliminating all the complementarity constraints $d_i^+ d_i^- = 0, i = 1, 2, \cdots, n+1$ and $y_t^+ y_t^- = 0, t = 1, 2, \cdots, T$, in (P3-7), we can get the following problem

$$\text{(P3-8)} \quad \max f(x) = \sum_{j=1}^{n+1} r_i x_i - x_{n+2}$$

$$\text{s.t.} \quad y_t^- \le w_0, \quad t = 1, 2, \cdots, T,$$

$$\sum_{i=1}^{n+1} k_i(d_i^+ + d_i^-) \le x_{n+2},$$

$$d_i^+ - d_i^- = x_i - x_i^0, \quad i = 1, 2, \cdots, n+1,$$

$$y_t^+ - y_t^- = \sum_{i=1}^{n} (r_{it} - r_i)x_i, \quad t = 1, 2, \cdots, T,$$

$$\sum_{i=1}^{n+1} x_i = 1,$$

$$\sum_{j=1}^{n+1} \left(\frac{la_j + lb_j}{2} + \frac{\beta_j - \alpha_j}{6} \right) x_j \geq E(\hat{l}_0),$$

$$0 \leq x_i \leq u_i, \quad i = 1, 2, \cdots, n+1,$$

$$d_i^+ \geq 0, \quad d_i^- \geq 0, \quad i = 1, 2, \cdots, n+1,$$

$$y_t^+ \geq 0, \quad y_t^- \geq 0, \quad t = 1, 2, \cdots, T.$$

Theorem3.3   Let $(x_1^*, \cdots, x_{n+2}^*, d_1^{+*}, \cdots, d_{n+1}^{+*}, d_1^{-*}, \cdots, d_{n+1}^{-*}, y_1^{+*}, \cdots, y_T^{+*}, y_1^{-*}, \cdots, y_T^{-*})$ is an optimal solution of (P3-7), then $(x_1^*, \cdots, x_{n+2}^*, d_1^{+'}, \cdots, d_{n+1}^{+'}, d_1^{-'}, \cdots, d_{n+1}^{-'}, y_1^{+'}, \cdots, y_T^{+'}, y_1^{-'}, \cdots, y_T^{-'})$ is an optimal solution of (P3-8), where

$$d_i^{+'} = \begin{cases} d_i^{+*} - d_i^{-*}, & \text{if } d_i^{+*} > d_i^{-*} > 0, \\ 0, & \text{if } d_i^{-*} \geq d_i^{+*} > 0, \end{cases}$$

$$d_i^{-'} = \begin{cases} 0, & \text{if } d_i^{+*} > d_i^{-*} > 0, \\ d_i^{-*} - d_i^{+*}, & \text{if } d_i^{-*} \geq d_i^{+*} > 0, \end{cases}$$

$$\begin{pmatrix} d_i^{+'} \\ d_i^{-'} \end{pmatrix} = \begin{pmatrix} d_i^{+*} \\ d_i^{-*} \end{pmatrix}, \quad \text{if } d_i^{+*} d_i^{-*} = 0, \tag{3.19}$$

$$y_t^{+'} = \begin{cases} y_t^{+*} - y_t^{-*}, & \text{if } y_t^{+*} > y_t^{-*} > 0, \\ 0, & \text{if } y_t^{-*} \geq y_t^{+*} > 0, \end{cases}$$

$$y_t^{-'} = \begin{cases} 0, & \text{if } y_t^{+*} > y_t^{-*} > 0, \\ y_t^{-*} - y_t^{+*}, & \text{if } y_t^{-*} \geq y_t^{+*} > 0, \end{cases}$$

$$\begin{pmatrix} y_t^{+'} \\ y_t^{-'} \end{pmatrix} = \begin{pmatrix} y_t^{+*} \\ y_t^{-*} \end{pmatrix}, \quad \text{if } y_t^{+*} y_t^{-*} = 0.$$

Proof   Since the objective functions of (P3-7) and (P3-8) are the same, and it is obvious that each feasible solution of (P3-7) is a feasible solution of (P3-8) also, we only need to prove that $(x_1^*, \cdots, x_{n+2}^*, d_1^{+'}, \cdots, d_{n+1}^{+'}, d_1^{-'}, \cdots, d_{n+1}^{-'}, y_1^{+'}, \cdots, y_T^{+'}, y_1^{-'}, \cdots, y_T^{-'})$ is a feasible solution of (P2-7). It is easy to verify that the constraint conditions 3–8 are satisfied. Since $d_i^{+'}, d_i^{-'}, i = 1, 2, \cdots, n+1$, $y_t^{+'}, y_t^{-'}, t = 1, 2, \cdots, T$ and $(x_1^*, \cdots, x_{n+2}^*, d_1^{+*}, \cdots, d_{n+1}^{+*}, d_1^{-*}, \cdots, d_{n+1}^{-*}, y_1^{+*}, \cdots, y_T^{+*}, y_1^{-*}, \cdots, y_T^{-*})$ is feasible. Due to

$$d_i^{+'} + d_i^{-'} = \begin{cases} d_i^{+*} - d_i^{-*} \leq d_i^{+*} + d_i^{-*}, & \text{if } d_i^{+*} > d_i^{-*} > 0, \\ d_i^{-*} - d_i^{+*} \leq d_i^{+*} + d_i^{-*}, & \text{if } d_i^{-*} \geq d_i^{+*} > 0, \\ d_i^{+*} + d_i^{-*}, & \text{if } d_i^{-*} d_i^{+*} = 0 \end{cases}$$

and

$$\sum_{i=1}^{n+1} k_i(d_i^{+'} + d_i^{-'}) \le \sum_{i=1}^{n+1} k_i(d_i^{+*} + d_i^{-*}) \le x_{n+1}^*.$$

We can prove that constraint condition 2 of (P3-7) is satisfied. By (3.19), it is easy to verify that constraint condition 1 is satisfied too.

By Theorem 3.3, we can get the solution of (P3-7) by solving (P3-8).

Let $y = \max\{y_t^-, t = 1, 2, \cdots, T\}$, and eliminate y_t^+, we can get the following problem

$$\text{(P2-9)}\quad \max f(x) = \sum_{i=1}^{n+1} r_i x_i - x_{n+2}$$

$$\text{s.t.}\quad y \le w_0,$$

$$\sum_{i=1}^{n+1} k_i(d_i^+ + d_i^-) \le x_{n+2},$$

$$d_i^+ - d_i^- = x_i - x_i^0, \quad i = 1, 2, \cdots, n+1,$$

$$y + \sum_{i=1}^{n}(r_{it} - r_i)x_i \ge 0, \quad t = 1, 2, \cdots, T,$$

$$\sum_{i=1}^{n+1} x_i = 1,$$

$$\sum_{j=1}^{n+1} \left(\frac{la_j + lb_j}{2} + \frac{\beta_j - \alpha_j}{6}\right) x_j \ge E(\hat{l}_0),$$

$$0 \le x_i \le u_i, \quad i = 1, 2, \cdots, n+1,$$

$$d_i^+ \ge 0, \ d_i^- \ge 0, \quad i = 1, 2, \cdots, n+1,$$

$$y \ge 0.$$

Thus, (P3-4) is transformed into a standard linear programming problem (P3-9). In the same way, (P3-3) can be transformed into the following problem

(P3-10) $\max f(x) = \displaystyle\sum_{i=1}^{n+1} r_i x_i - x_{n+2}$

s.t. $\dfrac{1}{T} \displaystyle\sum_{t=1}^{T} y_t \leq w_0,$

$\displaystyle\sum_{i=1}^{n+1} k_i(d_i^+ + d_i^-) \leq x_{n+2},$

$d_i^+ - d_i^- = x_i - x_i^0, \quad i = 1, 2, \cdots, n+1,$

$y_t + \displaystyle\sum_{i=1}^{n}(r_{it} - r_i)x_i \geq 0, \quad t = 1, 2, \cdots, T,$

$\displaystyle\sum_{i=1}^{n+1} x_i = 1,$

$\displaystyle\sum_{j=1}^{n+1} \left(\dfrac{la_j + lb_j}{2} + \dfrac{\beta_j - \alpha_j}{6} \right) x_j \geq E(\hat{l}_0),$

$0 \leq x_i \leq u_i, \quad i = 1, 2, \cdots, n+1,$

$d_i^+ \geq 0, \quad d_i^- \geq 0, \quad i = 1, 2, \cdots, n+1,$

$y_t \geq 0, \quad t = 1, 2, \cdots, T.$

(P3-9) and (P3-10) are linear programming problems. A few linear programming algorithms, for example, the simplex method, can be used to solve it efficiently.

3.5 Numerical Example

Chinese securities markets comprise the Shanghai Stock Exchange, the Shenzhen Stock Exchange, the Hong Kong Stock Exchange and the Taipei Stock Exchange. In this section, we give an example to illustrate the model for portfolio selection proposed in this chapter. We suppose that an investor wants to choose twelve different types of stocks and a kind of risk-less asset from the Shanghai Stock Exchange for his investment. The names of the twelve kinds of stocks are given in Table 12.1.

Table 3.1. Name of Stocks

Handan Gangtie	Qilu Shihua	Shanghai Jichang
Wukuang Fazhan	Gezhouba	Jiangnan Zhonggong
Guangzhou Konggu	Qinghua Tongfang	Shanghai Jiche
Dongfang Hangkong	Dongfang Jituan	Diyibaihuo

The rate of transaction costs for risk-less assets is 0.142 5%. The rate of transaction costs for risky assets is 0.0055.

Since we assume that the future turnover rates of the securities are trapezoidal fuzzy numbers, we need to estimate the tolerance interval, left width and right width of the fuzzy numbers. In the real world of portfolio management, the investor can obtain values of these parameters by using the Delphi Method, based on experts' knowledge. In our example, based on historical data of the securities turnover rates, we adopt the frequency statistic method to estimate these parameters. In the following, we give the estimation method for the fuzzy turnover rates for Stock Guangzhou Konggu in detail. First, we use historical data (daily turnover rates from March, 2000 to April, 2003) to calculate the frequency of historical turnover rates. Fig 4 shows the frequency distribution of historical turnover rates for the stocks. Note that most of the historical turnover rates fall into the intervals [0.020, 0.030], [0.030, 0.040] and [0.040, 0.050]. We assume mean values 0.033 and 0.047 of historical turnover rates fall into intervals [0.020, 0.030] and [0.040, 0.050], as the left and the right endpoints of the tolerance intervals, respectively. Therefore, the tolerance interval of the fuzzy turnover rate is [0.033, 0.047]. By observing all the historical data, we use 0.006 and 0.194 as the minimum and the maximum possible values of uncertain turnover rates in the future. Thus, the left width is 0.027 and the right width is 0.147. The fuzzy turnover rate of Stock 1 is (0.033, 0.047, 0.027, 0.147). Using a similar method, we obtain the fuzzy turnover rates of all the 12 stocks. These are listed in Table 4.

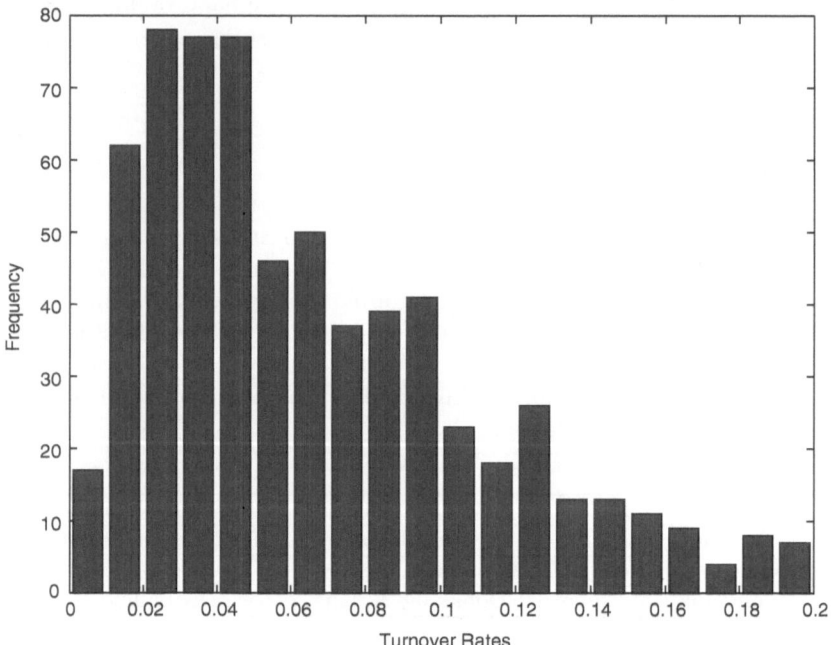

Fig. 3.1. Frequency distribution of historical turnover rates for StockGuangzhou Konggu

We collect historical data of the twelve stocks from January, 1999 to December, 2002. The data are downloaded from the web-site www.stockstar.com. Then we use one month as a period to obtain the historical rates of returns for forty-eight periods. With these historical data, the expected rates of returns of the stocks are listed in Table3.3.

Based on the above data, we can get investment strategies by applying the proposed models. Assume the minimum level of liquidity is 0.050. We can obtain the optimal investment strategy by solving (P3-9) or (P3-10). The risk and return of the portfolio obtained by (P3-9), are listed in Table3.4, the detailed investment is listed in Table3.5; risk and return of portfolio obtained by (P3-10) is listed in Table3.6, the detailed investment is listed in Table3.7. Based on the data in Table 3.4, we can get the efficient frontier of the mean-Minimax Semi-absolute deviation model (see Fig 3.2); based on the data in Table 3.6, we can get the efficient frontier of the mean-Semi-absolute deviation model (see Fig 3.3).

3.6 Conclusion

In addition to the more usual factors of expected return and risk, liquidity is considered in portfolio selection problems. The turnover rates of securities are used to measure their liquidity. Considering the three factors, a portfolio optimization model with fuzzy liquidity constraints is proposed. An example is given to illustrate the behavior of the proposed portfolio selection model, using real data from the Shanghai Stock Exchange. The computation results show that the portfolio selection model with fuzzy liquidity constraints can generate a favorite portfolio selection strategy, according to the investor's degree of satisfaction.

Table 3.2. Fuzzy turnover rate of stocks

Stock	tolerance	left width	right width
HDGT	[0.022, 0.034]	0.018	0.111
QLSH	[0.032, 0.044]	0.027	0.110
SHJC	[0.014, 0.025]	0.012	0.101
WKFZ	[0.016, 0.034]	0.012	0.082
GZB	[0.012, 0.026]	0.010	0.082
JNZG	[0.032, 0.096]	0.018	0.099
GZKG	[0.033, 0.047]	0.027	0.147
QHTF	[0.025, 0.037]	0.022	0.119
SHQC	[0.031, 0.057]	0.026	0.081
DFHK	[0.023, 0.046]	0.017	0.139
DFJT	[0.016, 0.045]	0.012	0.101
DYBH	[0.023, 0.036]	0.020	0.111

Table 3.3. Expected return of stocks

Stock	HDGT	QLSH	SHJC	WKFZ	GZB
Expected return	0.006 3	0.006 6	0.010 7	0.023 4	0.007 1
Stock	JNZG	GZKG	QHTF	SHQC	DFHK
Expected return	0.009 4	0.016 7	0.026 3	0.008 1	0.016 0
Stock	DFJT	DYBH			
Expected return	0.022 6	0.012 4			

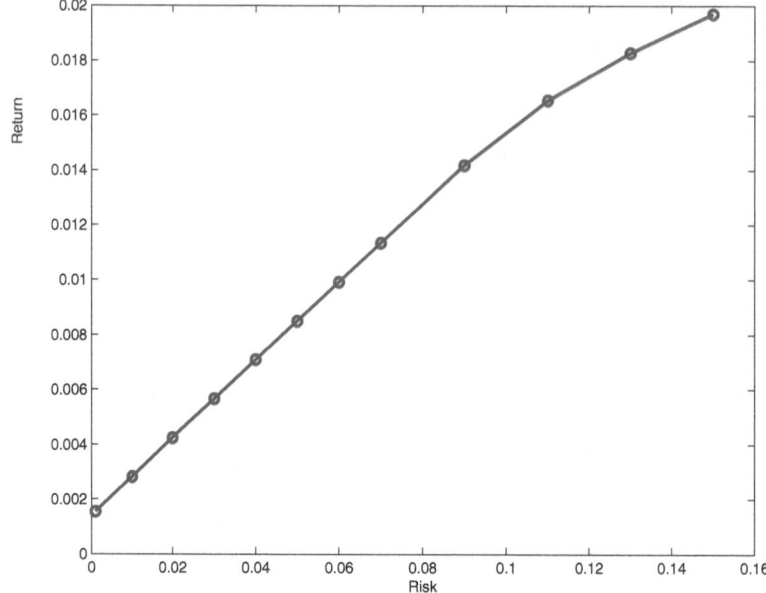

Fig. 3.2. Efficient frontier of Mean-Minimax Semi-absolute Deviation Model

Table 3.4. Risk, return and liquidity by (P2-9)

portfolio	1	2	3	4	5	6
return	0.001 54	0.002 82	0.004 24	0.005 66	0.007 08	0.008 50
risk	0.001	0.010	0.020	0.030	0.040	0.050
liquidity	0.099 5	0.095 2	0.090 4	0.085 7	0.080 9	0.076 1
portfolio	7	8	9	10	11	12
return	0.009 92	0.011 34	0.014 18	0.016 54	0.018 28	0.019 7
risk	0.060	0.070	0.090	0.110	0.130	0.150
liquidity	0.071 4	0.066 7	0.057 1	0.051 7	0.051 6	0.050 0

Table 3.5. Investment by (P3-9)

Stock	HDGT	QLSH	SHJC	WKFZ	GZB
portfolio 1	0.000 0	0.000 0	0.000 0	0.000 0	0.000 0
portfolio 4	0.000 0	0.000 0	0.000 0	0.000 0	0.000 0
portfolio 6	0.000 0	0.000 0	0.000 0	0.000 0	0.000 0
Stock	JNZG	GZKG	QHTF	SHQC	DFHK
portfolio 1	0.001 9	0.000 0	0.001 8	0.000 0	0.000 0
portfolio 4	0.056 0	0.000 0	0.055 3	0.000 0	0.000 0
portfolio 6	0.093 4	0.000 0	0.092 1	0.000 0	0.000 0
Stock	DFJT	DYBH	Saving		
portfolio 1	0.006 3	0.000 0	0.990 0		
portfolio 4	0.184 7	0.000 0	0.704 0		
portfolio 6	0.307 8	0.000 0	0.506 7		

Table 3.6. Risk, return and liquidity by (P3-10)

portfolio	1	2	3	4	5	6
return	0.001 93	0.002 46	0.004 06	0.006 72	0.009 38	0.012 05
risk	0.001	0.002	0.005	0.010	0.015	0.020
liquidity	0.098 2	0.096 5	0.091 4	0.082 8	0.074 2	0.065 6
portfolio	7	8	9	10	11	12
return	0.014 71	0.017 27	0.018 43	0.019 00	0.019 54	0.020 02
risk	0.025	0.030	0.035	0.040	0.045	0.050
liquidity	0.056 9	0.050 0	0.050 0	0.050 0	0.050 0	0.050 0

Table 3.7. Investment by (P3-10)

Stock	HDGT	QLSH	SHJC	WKFZ	GZB
portfolio 1	0.000 0	0.000 0	0.000 0	0.003 0	0.000 0
portfolio 8	0.000 0	0.000 0	0.000 0	0.046 0	0.000 0
portfolio 12	0.000 0	0.000 0	0.000 0	0.000 0	0.000 0
Stock	JNZG	GZKG	QHTF	SHQC	DFHK
portfolio 1	0.000 0	0.000 0	0.006 6	0.000 0	0.000 0
portfolio 8	0.000 0	0.000 0	0.267 7	0.000 0	0.000 0
portfolio 12	0.000 0	0.000 0	0.894 9	0.000 0	0.000 0
Stock	DFJT	DYBH	Saving		
portfolio 1	0.021 6	0.000 0	0.968 8		
portfolio 8	0.050 2	0.000 0	0.083 8		
portfolio 12	0.120 0	0.000 0	0.054 9		

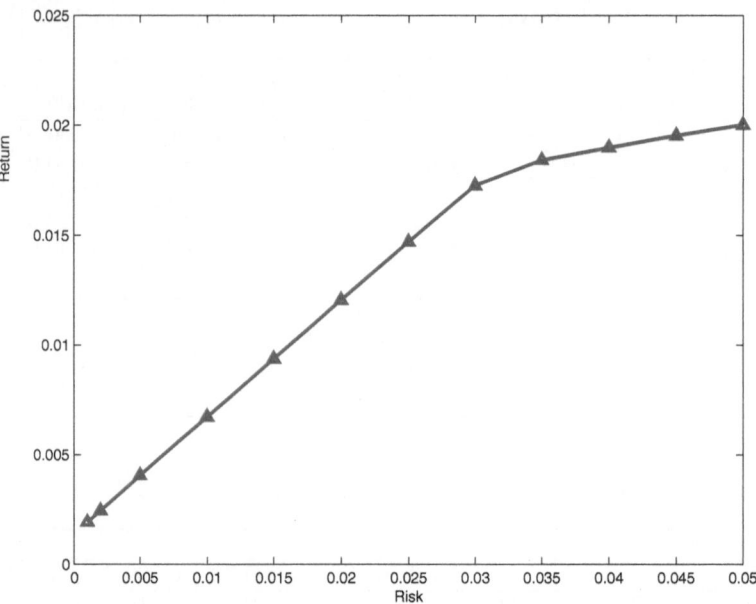

Fig. 3.3. Efficient frontier of Mean-Semi-absolute Deviation Model

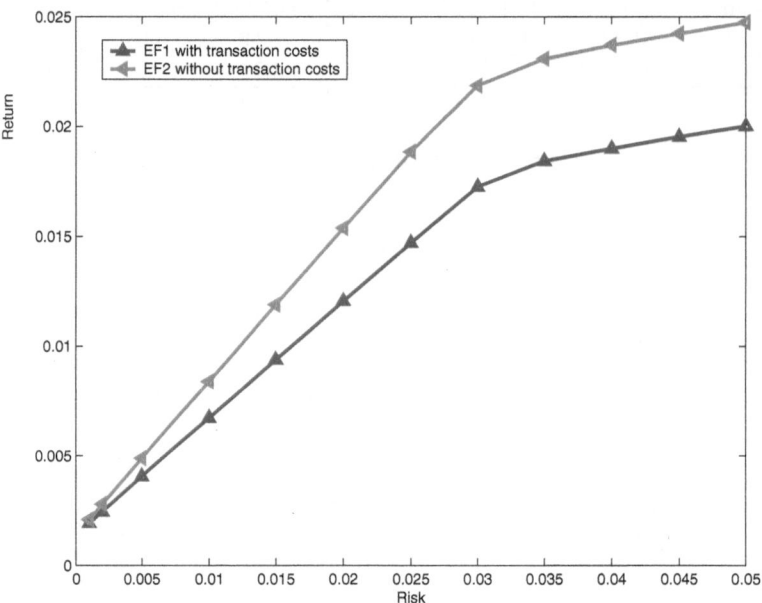

Fig. 3.4. The effect of transaction costs on the efficient frontier

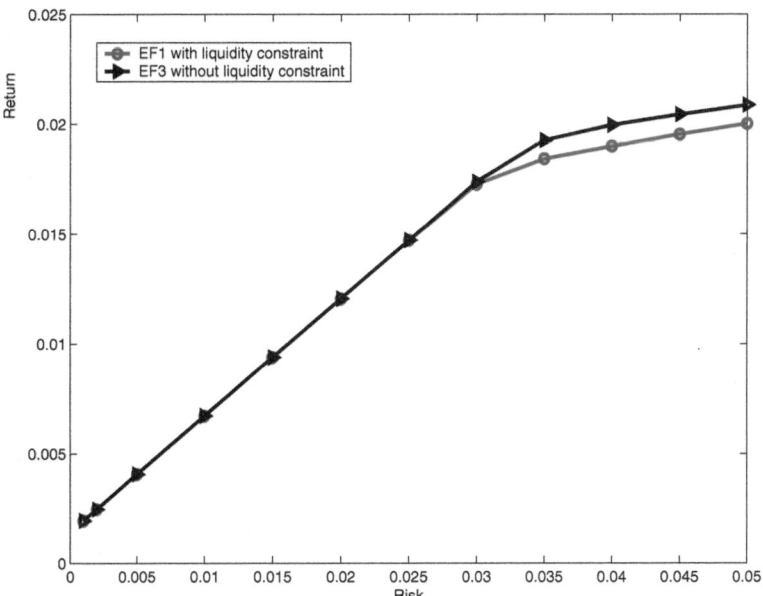

Fig. 3.5. The effect of liquidity on the efficient frontier

4

Ramaswamy's Model

4.1 Introduction

Ramaswamy (1998) presented a bond portfolio selection method using the fuzzy decision theory in a BIS (Bank for international settlements) working paper. The proposed approach can ensure that investors get a given minimum rate if return.

In order to meet the return target for assets under management, fund managers have to constantly judge the direction of financial market moves. Due to the inherent uncertainty of financial market, fund managers are very cautious in expressing their views about the market. Ramaswamy described the information content in such cautious views as fuzzy or vague, in terms of both the direction and the size of market moves. Ramaswamy assumed that the investment horizon is typically one to three months and the investment universe consists of government debt securities and plain vanilla options on these securities in his paper. The target rate of return were assumed to be a certain number of basis points above Libor over the given investment horizon. Considering the investment horizon and the selected securities, the methodology is suitable for central banks managing their short-term liquidity portfolio especially.

Ramaswamy pointed out that a fund manager structuring a fixed-income portfolio may have only vague views regarding future interest rate scenarios and these can broadly be described as being bullish, bearish or neutral. The vague views are based on decision maker's intuitive opinions. Under these fuzzy uncertain circumstances, the decision maker could characterize the range of acceptable solutions to the portfolio selection problem as a fuzzy set. Fuzzy set theory is a very efficient tool when dealing with problems in which the source of imprecision is the absence of sharply defined criteria of class membership rather than the presence of random variables. In next section, we will introduce the fixed-income portfolio selection problem in the framework of fuzzy decision theory.

4.2 Model Formulation

Assume that a fund manager who has to choose a structured portfolio from an investment universe of n assets with $x_i^{\min}$ and $x_i^{\max}$ being the minimum and maximum weight of asset i $(i = 1, 2, \cdots, n)$ in the portfolio. In order to select the structured portfolio, the fund manager may examine m potential market scenarios, and for each of these scenarios he/she may wish to maximize the portfolio return. R_{ik} denotes the return from asset i $(i = 1, 2, \cdots, n)$ in market scenario k $(k = 1, 2, \cdots, m)$ at the end of the investment period. So $R_k(x) = \sum_{i=1}^{n} R_{ik} x_i$ denote the portfolio return for scenario k $(k = 1, 2, \cdots, m)$.

Denote by $R_k^{\min}$ and $R_k^{\max}$ the minimum and the maximum expected return for market scenario k $(k = 1, 2, \cdots, m)$. Note that it is quite easy for the fund manager to provide information on the expected target range of return for various scenarios, rather than to define the a priori probabilities for different scenarios. Using the linear membership function $\mu_{R_k}(x)$, it is possible to compute the degree of satisfaction for any given portfolio x for market scenario k $(k = 1, 2, \cdots, m)$.

$$
\mu_{R_k}(x) = \begin{cases} 0, & \text{if} R_k(x) \leq R_k^{\min}, \\ \frac{R_k(x) - R_k^{\min}}{R_k^{\max} - R_k^{\min}}, & \text{if} R_k^{\min} < R_k(x) \leq R_k^{\max}, \\ 1, & \text{if} R_k(x) > R_k^{\max}. \end{cases}
$$

To achieve the return objective the fund manager could formulate the following optimization problem:

$$
\max \quad \min \{\mu_{R_1}(x), \mu_{R_2}(x), \cdots, \mu_{R_m}(x)\}
$$
$$
\text{s.t.} \quad b_j^{\min} \leq \sum_{i=1}^{n} a_{ij} x_i \leq b_j^{\max}, \quad j = 1, \cdots, P.
$$

By introducing λ, the above optimization problem can be transformed into the following problem

$$
\max \quad \lambda
$$
$$
\text{s.t.} \quad \mu_{R_k}(x) \geq \lambda, \quad k = 1, 2, \cdots, m,
$$
$$
b_j^{\min} \leq \sum_{i=1}^{N} a_{ij} x_i \leq b_j^{\max}, \quad j = 1, \cdots, P.
$$

For obtaining the goal of portfolio return, the fund manager can get investment strategy by solving the following problem

$$
\max \quad R_k = \sum_{i=1}^{n} R_{ik} x_i, \quad k = 1, 2, \cdots, m
$$
$$
\text{s.t.} \quad \sum_{i=1}^{n} x_i = 1,
$$
$$
x_i^{\min} \leq x_i \leq x_i^{\max}, \quad i = 1, \cdots, n.
$$

The above problem is a multi-objective programming problem; one can get a Pareto optimal solution by solving the following problem

$$
\begin{aligned}
\max \quad & \lambda \\
\text{s.t.} \quad & w_k R_k \geq \lambda, \quad k = 1, 2, \cdots, m, \\
& \sum_{i=1}^{n} x_i = 1, \\
& x_i^{\min} \leq x_i \leq x_i^{\max}, \quad i = 1, \cdots, n.
\end{aligned}
$$

Thus, Ramaswamy proposed the following portfolio selection model based on the fuzzy decision making theory

$$
\begin{aligned}
\max_{x, \lambda} \quad & \lambda \\
\text{s.t.} \quad & \mu_{R_k}(x) \geq \lambda, \quad k = 1, \cdots, m, \\
& \sum_{i=1}^{n} x_i = 1, \\
& x_i^{\min} \leq x_i \leq x_i^{\max}, \quad i = 1, \cdots, n.
\end{aligned}
$$

4.3 Conclusion

In this chapter, we introduce Ramaswamy's Model. Ramaswamy (1998) gave a numerical example in which the investor is only allowed to hold government bonds and plain vanilla options, and only two scenarios are assumed: "bullish" and "bearish". One can refer to Ramaswamy (1998) for details of the numerical example.

5

León-Liern-Vercher's Model

5.1 Formulations of Portfolio Selection Problem

Markowitz (1952) proposed the mean variance model for the portfolio selection problem. The model is a quadratic programming problem (MV), in which risk variance is minimized and investment diversification is treated in computational terms

$$(\text{MV}) \quad \min \sum_{i=1}^{n} \sum_{j=1}^{n} \sigma_{ij} x_i x_j$$

$$\text{s.t.} \sum_{i=1}^{n} E(R_i) x_i \geq \rho,$$

$$\sum_{i=1}^{n} x_i = 1,$$

$$l_i \leq x_i \leq u_i, \quad i = 1, \cdots, n,$$

where x_i represents the percentage of money invested in asset i, R_i is the random variable representing the return of asset i, σ_{ij} is the covariance between returns of asset i and of asset j, and ρ is a parameter representing the minimal rate of return required by an investor. Also, $u_i(l_i)$ is the maximum (minimum) amount of money which can be invested in asset i.

The average vector of returns, and the elements of the covariance matrix over T periods, can be approximated by

$$E(R_i) = \frac{1}{T} \sum_{k=1}^{T} r_{ik}, \quad i = 1, 2, \cdots, n$$

and

$$\hat{\sigma}_{ij} = \frac{1}{T} \sum_{k=1}^{T} (r_{ik} - E(R_i))(r_{jk} - E(R_j)), \quad i, j = 1, 2, \cdots, n,$$

where r_{ik} is the realization of the random variable R_i during period k ($k = 1, 2, \cdots, T$) and is obtainable through historical data.

Sharpe (1963) formulated the MV problem as a simplified quadratic programming model by using market indices to express assets' returns. The simplified quadratic model is called the single index model. It is well known that the mean variance portfolio models proposed by Markowitz gave rise to a variety of regression models, including the extensively used CAPM, which was subsequently developed by Sharpe and Lintner.

Konno and Yamazaki (1991) proposed a linear portfolio optimization model: the L_1 risk model. The measure of risk is minimizing the sum of absolute deviations from the averages associated with x_i choices. By using the same notation as in the (MV) problem, we have

$$\min E\left(\left|\sum_{i=1}^{n} R_i x_i - E\left(\sum_{i=1}^{n} R_i x_i\right)\right|\right)$$

$$\text{s.t. } \sum_{i=1}^{n} E(R_i)x_i \geq \rho M_0,$$

$$\sum_{i=1}^{n} x_i = M_0,$$

$$l_i \leq x_i \leq u_i, \quad i = 1, \cdots, n.$$

The above problem can be transformed into

$$\text{(LMAD)} \quad \min \frac{1}{T} \sum_{k=1}^{T} y_k$$

$$\text{s.t. } y_k + \sum_{i=1}^{n} (r_{ik} - E(\hat{R}_i))x_i \geq 0, \quad k = 1, 2, \cdots, T,$$

$$y_k - \sum_{i=1}^{n} (r_{ik} - E(\hat{R}_i))x_i \geq 0, \quad k = 1, 2, \cdots, T,$$

$$\sum_{i=1}^{n} E(\hat{R}_i))x_i \geq \rho M_0,$$

$$\sum_{i=1}^{n} x_i = M_0,$$

$$l_i \leq x_i \leq u_i, \quad i = 1, \cdots, n.$$

It has been shown that the MV and LMAD models usually generate similar portfolios. If the expected return, ρ, belongs to

$$\left[\min_{1 \leq i \leq n} \{E(R_i)\}, \max_{1 \leq i \leq n} \{E(R_i)\}\right],$$

and there are no diversification conditions (bound-type constraints on the assets) it is well known that the (MV) and (LMAD) problems are always feasible. But in many situations, when attempting to reflect the diversification proposed by the investor, infeasibility surfaces.

León, Liern and Vercher (2002) proposed an algorithm to repair infeasibility by using a fuzzy method. Results of the numerical example show that the conditions in this new feasible instance are valid and reasonable for the investor.

5.2 Analysis of Infeasibility of Portfolio Selection Problem

Denote the linearly constrained portfolio selection problem by

$$(\mathcal{P}) \quad \min\{f(x) : \mathcal{A}_1 x \geq \mathcal{B}^1, \mathcal{A}_2 x \leq \mathcal{B}^2, \mathcal{A}_3 x = \mathcal{B}^3, x \geq 0\},$$

where $x \in R^n, \mathcal{A}_i \in M_{p_i \times n}(R), \mathcal{B}^i \in R^{p_i}$ $(i = 1, 2, 3)$, f is a real-valued function in R^n, and assume that $\mathcal{P}$ is an infeasible instance. A conceptual assumption underlying León, Liern and Vercher's approach is that problem $\mathcal{P}$ is correctly formulated. In particular, assume that the investor diversification conditions are logical in the sense that the set of bounds verifies that

$$\{x \in R^n : l_i \leq x_i \leq u_i, 1 \leq i \leq n, \sum_{i=1}^n x_i = 1\} \neq \Phi.$$

It would be natural for every reasonable investor to provide a set of bounds like this. When the expected benefit is not compatible with the diversification constraints, the infeasibility appears. Then, removing or changing one single constraint (that associated to q, for instance) is not specially attractive.

Denote by X the set of hard constraints, i.e.

$$X := \{x \in R^n : H_1 x \geq \beta^1, H_2 x \leq \beta^2, \mathcal{A}_3 x = \mathcal{B}^3, x \geq 0\},$$

where $H_i \in M_{(p_i - m_i) \times n}(R), \beta^i \in R^{p_i - m_i}$ $(i = 1, 2)$ and $\mathcal{A}_3 \in M_{p_3 \times n}(R)$, $\mathcal{B}^3 \in R^{p_3}$.

Assume X is a nonempty set, then $\mathcal{P}$ can be rewritten as

$$(\mathcal{P}) \quad \begin{aligned} &\min f(x) \\ &\text{s.t. } A_1 x \geq b^1, \\ &\qquad A_2 x \leq b^2, \\ &\qquad x \in X, \end{aligned}$$

where $A_1 \in M_{m_1 \times n}(R), A_2 \in M_{m_2 \times n}(R), b^1 \in R^{m_1}, b^2 \in R^{m_2}$.

In order to attain feasibility in $\mathcal{P}$, León, Liern and Vercher relax these constraints in a certain degree. Assume that m_1 greater or equal type fuzzy constraints $\tilde{B}_1, \tilde{B}_2, \cdots, \tilde{B}_{m_1}$, m_2 less or equal type fuzzy constraints

$\tilde{C}_1, \tilde{C}_2, \cdots, \tilde{C}_{m_2}$, $m_i \leq p_i$ ($i = 1, 2$). Denote by $\mu_{\tilde{B}_i}(x)$ and $\mu_{\tilde{C}_i}(x)$ the membership functions for $\tilde{B}_i$ and $\tilde{C}_i$, respectively. León, Liern and Vercher gave the following concept of feasibility:

Definition 5.1 Define the fuzzy set of feasible solutions of $\mathcal{P}$ $\tilde{E} := \{(x, \mu_{\tilde{E}}(x)), x \in X\}$, where $\mu_{\tilde{E}}(x)$

$$\mu_{\tilde{E}}(x) = \min\{\mu_{\tilde{B}_1}(x), \cdots, \mu_{\tilde{B}_{m_1}}(x), \mu_{\tilde{C}_1}(x), \cdots, \mu_{\tilde{C}_{m_2}}(x)\}.$$

Notice that the fuzzy set $\tilde{E}$ is non-empty, so the following fuzzy program is consistent: (FP)   find $\{x \in X : A_1 x \geq b^1, A_2 \leq b^2\}$.

The solution with the highest degree of membership is given by

$$x_{\max} = \arg(\max_{x \in X} \min_{i,j}\{\mu_{\tilde{B}_i}(x), \mu_{\tilde{C}_j}(x)\}).$$

If the investor accepts the solution $x_{\max}$, ones have a viable portfolio selection. In the proposed approach, the objective function does not intervene in the proposed selection associated with a degree of investor satisfaction with respect to the constraints. We can find that if a decision maker also has aspiration levels for the risk, a symmetric fuzzy multi-objective formulation that no longer distinguishes between objectives and constraints can be used.

5.3 Fuzzy Portfolio Selection Model

Assume that the decision maker has target values for both the expected return rate ρ and the diversification conditions (l_i, u_i) and the targets lead to an infeasible instance of the problem. León, Liern and Vercher used an interactive system to solve the problem of getting a viable portfolio selection.

At the first stage, one can apply the scheme introduced in above section, which associates certain related membership functions to the soft inequality constraints while leaving the hard constraints, unchanged. At the second stage, investor's opinion can be used to select a portfolio in the framework of the trade-off analysis.

If one use $(Ax)_i \geq b_i, x \in R^n$ to represent a fuzzy inequality relation, the degree of satisfaction of this ith constraint is

$$\mu_{\tilde{B}_i}(x) = \begin{cases} 0, & \text{if } (Ax)_i < b_i - r_i, \\ g_i((Ax)_i), & \text{if } b_i - r_i \leq (Ax)_i \leq b_i, \\ 1, & \text{if } (Ax)_i \geq b_i, \end{cases}$$

where r_i is the maximum violation allowed for the ith constraint and it is usually assumed that $g_i((Ax)_i) \in [0, 1]$ is such that the higher the violation of the constraint, the lower the value of $g_i((Ax)_i)$.

Analogously for the fuzzy relation $(Ax)_i \leq b_i, x \in R^n$ the membership function is

$$\mu_{\tilde{C}_j}(x) = \begin{cases} 0, & \text{if } (Ax)_j > b_j + s_j, \\ g_j((Ax)_j), & \text{if } b_j \leq (Ax)_j \leq b_j + s_j, \\ 1, & \text{if } (Ax)_j \leq b_j. \end{cases}$$

Several types of function $g_i((Ax)_i)$ and $g_j((Ax)_j)$ can be used to construct the membership functions for the fuzzy constraints. Then, the decision maker provides or the analyst determines the tolerances (r_i, s_j). One can compute the tolerances by means of the shadow prices of the solution of the Phase I problem (PI), associated to the infeasible instance P:

$$\text{(PI)} \quad \min \sum_{i=1}^{m_1} a_i$$

$$\begin{aligned} \text{s.t.} \quad & A_1 x - I^{m_1} h + I^{m'} h + I^{m_1} a = b^1(\omega), \\ & A_2 x + I^{m_2} h' = b^2(\pi), \\ & H_1(x) \geq \beta^1(q), \\ & H_2(x) \leq \beta^2(p), \\ & A_3 x = B^3(y), \\ & h, h', a \geq 0, x \in X. \end{aligned}$$

where h are the slack variables, and a are the artificial ones. The dual variables associated to the soft constraints are ω and π and the dual variables corresponding to the hard constraints are denoted by q, p and y.

Let z^* be the optimal value of (PI) that provides the sum of infeasibilities and let $(\omega^*, \pi^*, q^*, p^*, y^*)$ be an optimal solution of its dual. The tolerances are defined by:

$$r_i = \begin{cases} 0, & \text{if } \omega_i^* = 0, \\ \frac{z^*}{\omega_i^*}, & \text{if } \omega_i^* > 0. \end{cases}$$

$$s_j = \begin{cases} 0, & \text{if } \pi_i^* = 0, \\ -\frac{z^*}{\pi_i^*}, & \text{if } \pi_i^* < 0. \end{cases}$$

For each soft constraint, ones have both the goal and the tolerance of the investor. ones can therefore construct their membership functions, except for those with nil tolerance, which do not need to be perturbed. Concerning membership functions for the soft diversification constraints, $x_i \geq l_i$ or $x_j \leq u_j$ for some i; j, León, Liern and Vercher propose to use linear functions, i.e.,

$$\mu_{\tilde{B}_i}(x) = g_i((Ax)_i) = 1 - \frac{l_i - x_i}{r_i}, \quad l_i - r_i < x_i < l_i$$

and

$$\mu_{\tilde{C}_j}(x) = g_j((Ax)_j) = 1 - \frac{x_j - u_j}{s_j}, \quad u_j < x_j < u_j + s_j.$$

Denote the expected return constraint,

$$\sum_{i=1}^{n} E(R_i) x_i \geq \rho.$$

. Ones will also use a linear membership function for this constraint. In case the decision maker should want to assign higher preference to the expected return than the diversification goals, another type of membership function should be used; but this must be done at the second stage of the proposed algorithm.

León, Liern and Vercher proposed a procedure to make an infeasible instance of P viable. At the first stage, a feasible portfolio selection is determined; then ones ask for investor opinion in order to modify the proposal, if required.

Stage I: Viability

Step 1: Classifying the constraints. Firstly determining which soft and hard constraints are prescribed in the linearly constrained problem P. They could be different depending on the considering model (MV, LMAD, etc.).

Step 2: Computing the tolerances and defining the membership functions.

Let z^* be the optimal value of (PI), and let $(\omega_1^*, \cdots, \omega_{m_1}^*, \pi_1^*, \cdots, \pi_{m_2}^*)$ be an optimal solution of its dual. Calculate the tolerances and the vectors: $R = (r_1, r_2, \cdots, r_n)$ and $S = (s_1, s_2, \cdots, s_n)$, where

$$r_i = \begin{cases} 0, & \text{if } \omega_i^* = 0, \\ \frac{z^*}{\omega_i^*}, & \text{if } \omega_i^* > 0. \end{cases}$$

$$s_j = \begin{cases} 0, & \text{if } \pi_i^* = 0, \\ -\frac{z^*}{\pi_i^*}, & \text{if } \pi_i^* < 0. \end{cases}$$

As it will be shown in Theorem 1, if the user considers it acceptable to modify the RHS terms of the soft constraints by at least $1/k$ in the direction (R, S), where k is the number of non-null components of vector (R, S), it makes sense to construct the membership functions in order to obtain a viable instance. A linear membership function for every soft diversification constraint with a non-null shadow price is considered.

$$\mu_{\tilde{B}_i}(x) = g_i((Ax)_i) = 1 - \frac{l_i - x_i}{r_i}, \quad l_i - r_i < x_i < l_i$$

and

$$\mu_{\tilde{C}_j}(x) = g_j((Ax)_j) = 1 - \frac{x_j - u_j}{s_j}, \quad u_j < x_j < u_j + s_j.$$

Step 3: Auxiliary linear problem. In order to determine the best solution, i.e. the solution with the highest degree of satisfaction in $\tilde{E}$, the following auxiliary crisp problem is need to solve:

$$\begin{aligned} \text{(AP)} \quad & \max \phi \\ & \text{s.t.} \quad A_1 x + \phi R \geq b^1, \\ & \qquad A_2 x - \phi S \leq b^2, \\ & \qquad x \in X. \end{aligned}$$

Let $(x_{\max}, \phi_{\min})$ be the optimal solution of (AP), the degree of satisfaction of $x_{\max}$ is $\lambda^* = 1 - \phi^*$ where $\phi^* = \min\{\phi_{\min}, 1\}$.

Step 4: Solving the fuzzy reformulation. Applying a parametric programming formulation for the resource set to the models,

$$\begin{aligned}
\text{(FP)} \quad \min \ & f(x) \\
\text{s.t.} \ & A_1 x + \phi R \geq b^1, \\
& A_2 x - \phi S \leq b^2, \\
& x \in X.
\end{aligned}$$

and then one can obtain a set of solutions $\phi^* \leq \phi \leq 1$, $x(\phi) = \text{Argmin}\{f(x) : A_1 x \geq b^1 - \phi R, \ A_2 x \leq b^2 + \phi S, \ x \in X\}$ depending on the values of parameters ϕ, and $\phi^* \leq \phi \leq 1$.

Stage II: Investor opinion

Option 1: Satisfying solution. One can ask the decision maker for reasonable values of parameter ϕ and solve the corresponding crisp problem (MV or LMAD). Notice that for any solution x_0^*, obtained for a given ϕ_0, its degree of satisfaction is $\lambda_0 = 1 - \phi_0$.

Option 2: Non-satisfying solution.

(a) Because of the constraints, suppose that at the original stage one had the condition $x_{i0} \leq u_{i0}$, and that the new values proposed at the end of Stage I, $[u_{i0}, \tilde{u}_{i0}]$, do not seem appropriate to the decision maker, then choosing a quantity u'_{i0}. Then fixing $x_{i0} = u'_{i0}$, and substituting it into the model. These arguments are clearly extensible to cases involving more than one constraint. These new modifications could provoke other disagreements.

(b) Because of the risk in this case, one think that a fuzzy multi-objective decision approach would be the most appropriate. The following results justify that the algorithm is well defined.

Theorem5.1 If problem (AP) is feasible, then $\phi_{\min} \geq 1/k$, where k denotes the number of non-null components of vector (R, S).

Proof. It suffices to construct a feasible solution for the dual of (AP) with objective value $1/k$, because this provides us with a lower bound for the optimal value of (AP).

Considering the dual problem of (PI):

$$\begin{aligned}
\text{(DPI)} \quad \max \ & \omega b^1 + \pi b^2 + q\beta^1 + p\beta^2 + y\mathcal{B}^3 \\
\text{s.t.} \ & \omega A_1 + \pi A_2 + q H_1 + p H_2 + y\mathcal{A}_3 \leq 0, \\
& 0 \leq \omega_i \leq 1, \ \pi_j \leq 0, \ \forall i, j, \\
& q \geq 0, \ p \leq 0.
\end{aligned}$$

whose optimal solution was denoted by $(\omega^*, \pi^*, q^*, p^*, y^*)$ and its optimal value as z^*.

And considering the dual problem of (AP):

(DAP)   max $\gamma b^1 + \tau b^2 + g\beta^1 + t\beta^2 + v\mathcal{B}^3$

s.t. $\gamma A_1 + \tau A_2 + gH_1 + tH_2 + v\mathcal{A}_3 \leq 0,$

$\gamma R - \tau S \leq I,$

$\gamma \geq 0, \quad \tau \leq 0,$

$g \geq 0, \quad t \leq 0.$

Fixing

$$\gamma = \frac{\omega^*}{kz^*}, \quad \tau = \frac{\pi^*}{kz^*}, \quad g = \frac{q^*}{kz^*}, \quad t = \frac{p^*}{kz^*}, \quad v = \frac{y^*}{kz^*},$$

one has a feasible solution (γ, τ, g, t, v) for (DAP) with objective value $1/k$.

Corollary 5.1 Given an infeasible instance of the portfolio selection problem $(\mathcal{P})$, our algorithm obtains a solution with satisfaction degree $\lambda^* \in [0, 1 - 1/k]$.

5.4 Numerical Example

In order to show the performance of our method, León, Liern and Vercher use the set of historical data shown in Table 1, used by Markowitz in 1959. The columns 2-10 represent American Tobacco, A.T.& T., United States Steel, General Motors, Atcheson & Topeka & Santa Fe, Coca-Cola, Borden, Firestone and Sharon Steel securities data.

Suppose that an investor wants to allocate one unit of wealth among each of the nine assets. Their expected return q must be greater than or equal to 16.5%. The portfolio must be selected in such a way that the minimum investments in assets 1, 3 and 6 must be of 5%, 7.5% and 7.5% of the total, respectively, and the maximum investments in assets 4 and 5 will be 33% and 25% of the total, i.e.,

$$l_1 = 0.05, \quad l_3 = l_6 = 0.075, \quad u_4 = 0.33, \quad u_5 = 0.25.$$

So $\mathcal{P}$ is an infeasible instance. The Phase I linear program associated to $\mathcal{P}$.

(PI)   min $a_1 + a_3 + a_6 + a_1'$

s.t. $E(x) + a_1' - h_1' = \rho,$

$x_i - h_i + a_i = l_i, \quad i = 1, 3, 6,$

$x_j + h_j = u_j, \quad j = 4, 5,$

$$\sum_{i=1}^{9} x_i = 1,$$

$x_i, h_i, h_i', a_i, a_1' \geq 0, \quad i = 1, \cdots, 9.$

The optimal value of (PI) is $z^* = 0.771\ 861 \times 10^{-2}$, and the dual prices are

$$\omega_1 = 1, \quad \omega_2 = 0.080\ 111, \quad \omega_3 = 0,$$

Table 5.1. Historical returns of securities 1

	1 Am.T	2 A.T.T	3 U.S.S	4 G.M.	5 A.T.Sfe
1937	−0.305	−0.173	−0.318	−0.477	−0.457
1938	0.513	0.098	0.285	0.714	0.107
1939	0.055	0.200	−0.047	0.165	−0.424
1940	−0.126	0.03	0.104	−0.043	−0.189
1941	-0.280	−0.183	−0.171	−0.277	0.637
1942	−0.003	0.067	−0.039	0.476	0.865
1943	0.428	0.300	0.149	0.225	0.313
1944	0.192	0.103	0.260	0.290	0.637
1945	0.446	0.216	0.419	0.216	0.373
1946	−0.088	−0.046	−0.078	−0.272	−0.037
1947	−0.127	−0.071	0.169	0.144	0.026
1948	−0.015	0.056	−0.035	0.107	0.153
1949	0.305	0.038	0.133	0.321	0.067
1950	−0.096	0.089	0.732	0.305	0.579
1951	0.016	0.090	0.021	0.195	0.040
1952	0.128	0.083	0.131	0.390	0.434
1953	−0.010	0.035	0.006	−0.072	−0.027
1954	0.154	0.176	0.908	0.715	0.469

$$\omega_4 = 0.090\ 944, \quad \pi_1 = -0.027\ 389, \quad \pi_2 = -0.052\ 056.$$

Then, the tolerances that define the vectors R, S are:

$$R = (0.007\ 719, 0.096\ 349, 0, 0.084\ 872),$$

$$S = (0.281\ 815, 0.148\ 276).$$

The optimal value of the associated auxiliary problem (AP) is $\phi_{\min} = 1/5$, the lower bound in 5.3. If the investor accepts this proposal, the new RHS values appear in Table 5.4. These results are valid both for MV and LMAD objectives, because the objective function has not intervened until now. Table 5.5 and Table 5.6 show the portfolio selection obtained by applying the parametric programming formulation, for both MV and LMAD models.

Suppose that the investor wants to reduce the risk associated to the portfolio with satisfaction 0.8 by approximately 10%, i.e. desired risk is 0.058. However, it is more important not to decrease (too much) the expected return. As the risk value associated with 0.7 is lower than DS, a portfolio selection with a satisfaction level greater than 0.7 and lower than 0.8 could be determined by considering a grid for $\lambda \in [0.7, 0.8]$. As the investor considers the expected benefits constraint more important than the remaining constraints, one can consider the risk as a soft constraint, with a non-linear membership function.

León, Liern and Vercher calculate the tolerances analogously in Step 2 (it is not exactly the same because in this case we have a non-linear constraint).

Table 5.2. Historical returns of securities 2

	6	7	8	9
	C.C	Bdn	Frstn.	S.S
1937	−0.065	−0.319	−0.400	−0.435
1938	0.238	0.076	0.336	0.238
1939	−0.078	0.381	−0.093	−0.295
1940	−0.077	−0.051	−0.090	−0.036
1941	−0.187	0.087	-y0.194	−0.240
1942	0.156	0.262	0.113	0.126
1943	0.351	0.341	0.580	0.639
1944	0.233	0.227	0.473	0.282
1945	0.349	0.352	0.229	0.578
1946	−0.209	0.153	−0.126	0.289
1947	0.355	−0.099	0.009	0.184
1948	−0.231	0.038	0.000	0.114
1949	0.246	0.273	0.223	−0.222
1950	−0.248	0.091	0.650	0.327
1951	−0.064	0.054	−0.131	0.333
1952	0.079	0.109	0.175	0.062
1953	0.067	0.210	−0.084	−0.048
1954	0.077	0.112	0.756	0.185

Table 5.3. Results of the procedure of viability

original values		transformed values
$\rho = 0.165$	$\tilde{\rho} = \rho - r_1 \phi^*$	= 0.163 456
$l_1 = 0.05$	$\tilde{l}_1 = 0.05 - r_2 \phi^*$	= 0.030 730
$l_3 = 0.075$	$\tilde{l}_3 = 0.075 - r_3 \phi^*$	= 0.075 000
$l_6 = 0.075$	$\tilde{l}_6 = 0.075 - r_4 \phi^*$	= 0.058 026
$u_4 = 0.33$	$\tilde{u}_4 = 0.33 + s_1 \phi^*$	= 0.386 363
$u_5 = 0.25$	$\tilde{u}_5 = 0.25 + s_2 \phi^*$	= 0.279 655

They add to (PI) the following constraint: $V(x) + h^* = DS$, where $V(x) = \sum_{i=1}^{n} \sum_{j=1}^{n} \sigma_{ij} x_i x_j$, h^* is a non-negative slack variable. The new shadow prices and tolerances are included in Table 5.7.

In order to introduce the decision maker's level of preference for the risk constraint one can use an exponential membership function, i.e.

$$\mu_{\tilde{V}}(x) = g(V(x)) = \frac{1 - \exp\left(\frac{-k(V^- - V(x))}{tol}\right)}{1 - \exp(-k)},$$

Table 5.4. Parametric portfolio selection 1

ϕ	0.2	0.3	0.4
Return	0.163 46	0.162 68	0.161 91
ObjMV	0.064 50	0.054 23	0.050 03
ObjLMAD	0.210 51	0.193 71	0.184 61
Investment			
x_1	0.030 73	0.021 10	0.011 46
x_2	0.000 00	0.000 00	0.000 00
x_3	0.245 23	0.075 00	0.075 00
x_4	0.386 36	0.388 83	0.322 79
x_5	0.279 66	0.294 48	0.309 31
x_6	0.058 03	0.049 54	0.041 05
x_7	0.000 00	0.171 05	0.240 39
x_8	0.000 00	0.000 00	0.000 00
x_9	0.000 00	0.000 00	0.000 00

Table 5.5. Parametric portfolio selection 2

ϕ	0.5	0.6	0.7
Return	0.161 14	0.160 37	0.159 60
ObjMV	0.046 67	0.044 75	0.043 20
ObjLMAD	0.175 51	0.169 79	0.165 93
Investment			
x_1	0.001 83	0.000 00	0.000 00
x_2	0.000 00	0.000 00	0.000 00
x_3	0.075 00	0.075 00	0.075 00
x_4	0.256 76	0.215 92	0.209 02
x_5	0.324 14	0.329 41	0.314 22
x_6	0.032 56	0.024 08	0.015 59
x_7	0.309 72	0.355 59	0.386 20
x_8	0.000 00	0.000 00	0.000 00
x_9	0.000 00	0.000 00	0.000 00

where $DS < V(x) < V^- = DS + tol$, $tol = 0.036\ 476$. Take $k = -5$, and solve the crisp problem:

$$\text{(MAP)} \quad \max \alpha$$
$$\text{s.t.} \quad V(x) - \frac{tol}{k}\ln\left(1 - \alpha(1 - \exp(-k))\right) \leq V^-,$$
$$A_1 x + (1 - \alpha)R \geq b^1,$$
$$A_2 x - (1 - \alpha)S \geq b^2,$$
$$x \in X.$$

Let (x^*, α^*) be the optimal solution of this non-linear programming problem. The degree of satisfaction of x^* is α^*. Table 5.8 shows the optimal

Table 5.6. Parametric portfolio selection 3

ϕ	0.8	0.9	1
Return	0.158 80	0.158 05	0.157 28
ObjMV	0.041 76	0.040 50	0.039 70
ObjLMAD	0.162 06	0.158 49	0.156 42
Investment			
x_1	0.000 00	0.000 00	0.000 00
x_2	0.000 00	0.000 00	0.000 00
x_3	0.075 00	0.075 00	0.075 00
x_4	0.202 12	0.195 55	0.190 64
x_5	0.299 03	0.285 05	0.277 29
x_6	0.007 10	0.000 00	0.000 00
x_7	0.416 75	0.444 40	0.457 06
x_8	0.000 00	0.000 00	0.000 00
x_9	0.000 00	0.000 00	0.000 00

Table 5.7. Tolerances of soft constraints

RHS prices	Shadow price	tolerance
$DS = 0.058\ 049$	$-0.226\ 865$	0.036 476
$\rho = 0.165$	1.0	0.008 275
$l_1 = 0.05$	0.068 907	0.120 092
$l_3 = 0.075$	0.0	0.0
$l_6 = 0.075$	0.074 182	0.111 553
$u_4 = 0.33$	$-0.023\ 036$	0.359 230
$u_5 = 0.25$	$-0.049\ 085$	0.168 588

portfolio selection associated to $\alpha^* = 0.804\ 713$, where $V(x^*) = 0.059\ 623$ and the expected return is 0.163384. They have reduced the risk by 8% and the expected benefit by less than 0.1%.

Table 5.8. Investment

$x_1 = 0.026\ 548$	$x_2 = 0.000\ 000$	$x_3 = 0.161\ 655$
$x_4 = 0.400\ 153$	$x_5 = 0.282\ 923$	$x_6 = 0.053\ 215$
$x_7 = 0.075\ 506$	$x_8 = 0.000\ 000$	$x_9 = 0.000\ 000$

5.5 Conclusion

In this chapter, we introduce León, Liern and Vercher's model (2002). They propose to use a specialized a fuzzy method that they have developed to repair infeasibility in linearly constrained problems. Their version takes into account the special structure of the constraints in linear and quadratic programming models for the portfolio selection problem, in such a way that the diversification and the expected return conditions are considered as soft constraints, while the remaining are hard constraints.

6

Fuzzy Semi-absolute Deviation Portfolio Rebalancing Model

6.1 Introduction

In 1952, Markowitz published his pioneering work which laid the foundation of modern portfolio analysis. Markowitz's model has served as a basis for development of the modern financial theory over the past five decades. However, contrary to its theoretical reputation, it is not used extensively to construct portfolios at a large-scale. One of the most important reasons for this is the computational difficulty associated with solving a large-scale quadratic programming problem with a dense covariance matrix. Konno and Yamazaki used the absolute deviation risk function to replace the risk function in Markowitz's model and formulated a mean absolute deviation portfolio optimization model. It turns out that the mean absolute deviation model maintains the favorable properties of Markowitz's model and removes most of the principal difficulties in solving Markowitz's model. Simaan provided a thorough comparison of the mean variance model and the mean absolute deviation model. Furthermore, Speranza used the semi-absolute deviation to measure the risk and formulated a portfolio selection model.

Transaction cost is one of the main concerns for portfolio managers. Arnott and Wagner found that ignoring transaction costs could result in an inefficient portfolio. Yoshimoto's empirical analysis also drew the same conclusion. Mao, Jacob, Patel and Subhmanyam and Morton and Pliska studied portfolio optimization with fixed transaction costs. Pogue, Chen, Jen and Zionts, and Yoshimoto studied portfolio optimization with variable transaction costs. Mulvey and Vladimirou and Dantzig and Infanger incorporated transaction costs into the multi-period portfolio selection model. Li, Wang and Deng gave a linear programming algorithm to solve a general mean variance model for portfolio selection with transaction costs. Behavior of the financial markets and investors' attitudes towards risks and returns have changed drastically over the past five decades and, therefore, most of portfolio optimization exercises involve a revision of an existing portfolio, *i.e.*, portfolio rebalancing.

In this study, we consider the liquidity angle, and we propose a linear programming model for portfolio rebalancing, after considering transaction costs; based on the fuzzy decision theory, a portfolio rebalancing model with transaction costs is proposed.

6.2 Linear Programming Model for Portfolio Rebalancing with Transaction Costs

Suppose an investor allocates his/her wealth among n securities offering random rates of return. The investor starts with an existing portfolio and decides how to reallocate assets.

The expected rate of return r_i of security i without transaction costs is given by

$$r_i = \frac{1}{T} \sum_{t=1}^{T} r_{it}, \ i = 1, 2, \cdots, n, \tag{6.1}$$

where r_{it} can be determined by historical or forecast data.

Let $x^+ = (x_1^+, x_2^+, \cdots, x_n^+)$ and $x^- = (x_1^-, x_2^-, \cdots, x_n^-)$, where x_i^+ is the proportion of security i, $i = 1, 2, \cdots, n$ bought by the investor, x_i^- is the proportion of security i, $i = 1, 2, \cdots, n$ sold by the investor. Then the transaction costs of security i, $i = 1, 2, \cdots, n$ can be expressed as

$$C_i(x_i^+, x_i^-) = p(x_i^+ + x_i^-), \quad i = 1, 2, \cdots, n, \tag{6.2}$$

where p is the rate of transaction costs for the securities. So the total transaction costs can be expressed as

$$C(x^+, x^-) = \sum_{i=1}^{n} p(x_i^+ + x_i^-). \tag{6.3}$$

We assume that the investor does not wish to invest additional capital in the portfolio rebalancing process. Thus, we have

$$\sum_{i=1}^{n} (x_i^0 + x_i^+ - x_i^-) + \sum_{i=1}^{n} p(x_i^+ + x_i^-) = 1. \tag{6.4}$$

where x_i^0 is the proportion of security i, $i = 1, 2, \cdots, n$ owned by the investor before portfolio rebalancing.

The expected net return on the portfolio after paying transaction costs is given by

$$\sum_{i=1}^{n} r_i(x_i^0 + x_i^+ - x_i^-) - \sum_{i=1}^{n} p(x_i^+ + x_i^-). \tag{6.5}$$

The semi-absolute deviation of return on portfolio $x = (x_1, x_2, \cdots, x_n)$ below the expected return over the past period t, $t = 1, 2, \cdots, T$ can be represented as

$$w_t(x) = |\min\{0, \sum_{i=1}^{n}(r_{it} - r_i)x_i\}| = \frac{|\sum_{i=1}^{n}(r_{it} - r_i)x_i| + \sum_{i=1}^{n}(r_i - r_{it})x_i}{2} \qquad (6.6)$$

where $x_i = x_i^0 + x_i^+ - x_i^-$.

So the expected semi-absolute deviation of the return on portfolio $x = (x_1, x_2, \cdots, x_n)$ below the expected return can be represented as

$$w(x) = \frac{1}{T}\sum_{t=1}^{T} w_t(x) = \sum_{t=1}^{T} \frac{|\sum_{i=1}^{n}(r_{it} - r_i)x_i| + \sum_{i=1}^{n}(r_i - r_{it})x_i}{2T} \qquad (6.7)$$

where $x_i = x_i^0 + x_i^+ - x_i^-$.

We use $w(x)$ to measure the portfolio risk.

A fuzzy number A is called trapezoidal with tolerance interval $[a, b]$, left width α and right width β if its membership function takes the following form:

$$A(t) = \begin{cases} 1 - \frac{a-t}{\alpha} & \text{if } a - \alpha \le t \le a, \\ 1 & \text{if } a \le t \le b, \\ 1 - \frac{t-b}{\beta} & \text{if } a \le t \le b + \beta, \\ 0 & \text{otherwise} \end{cases} \qquad (6.8)$$

and we denote $A = (a, b, \alpha, \beta)$. It can easily be shown that

$$[A]^\gamma = [a - (1 - \gamma)\alpha, b + (1 - \gamma)\beta], \forall \gamma \in [0, 1], \qquad (6.9)$$

where $[A]^\gamma$ denotes the γ-level set of A.

Let $[A]^\gamma = [a_1(\gamma), a_2(\gamma)]$ and $[B]^\gamma = [b_1(\gamma), b_2(\gamma)]$ be fuzzy numbers and let $k \in R$ be a real number. Using the extension principle we can verify the following rules for addition and scalar multiplication of fuzzy numbers:

$$[A + B]^\gamma = [a_1(\gamma) + b_1(\gamma), a_2(\gamma) + b_2(\gamma)], \qquad (6.10)$$

$$[kA]^\gamma = k[A]^\gamma. \qquad (6.11)$$

Carlsson and Fullér introduced the notation of crisp possibilistic mean value and crisp possibilistic variance of continuous possibility distributions, which are consistent with the extension principle. The crisp possibilistic mean value of A is

$$E(A) = \int_0^1 \gamma(a_1(\gamma) + a_2(\gamma))d\gamma. \qquad (6.12)$$

It is clear that if $A = (a, b, \alpha, \beta)$ is a trapezoidal fuzzy number, then

$$E(A) = \int_0^1 \gamma[a - (1 - \gamma)\alpha + b + (1 - \gamma)\beta]d\gamma = \frac{a+b}{2} + \frac{\beta - \alpha}{6} \qquad (6.13)$$

Denote the turnover rate of security j by trapezoidal fuzzy number $\hat{l}_j = (la_j, lb_j, \alpha_j, \beta_j)$. Then the turnover rate of portfolio $x = (x_1, x_2, \cdots, x_n)$ is $\sum_{j=1}^{n} \hat{l}_j x_j$.

By the definition, the crisp possibilistic mean value of the turnover rate of security j is represented as follows:

$$E(\hat{l}_j) = \int_0^1 \gamma[la_j - (1-\gamma)\alpha_j + lb_j + (1-\gamma)\beta_j]d\gamma = \frac{la_j + lb_j}{2} + \frac{\beta_j - \alpha_j}{6}. \quad (6.14)$$

Therefore, the crisp possibilistic mean value of the turnover rate of portfolio $x = (x_1, x_2, \cdots, x_n)$ can be represented as

$$E(\hat{l}(x)) = E(\sum_{j=1}^{n} \hat{l}_j x_j) = \sum_{j=1}^{n} (\frac{la_j + lb_j}{2} + \frac{\beta_j - \alpha_j}{6})x_j. \quad (6.15)$$

In the study, we use the crisp possibilistic mean value of the turnover rate to measure the portfolio liquidity.

Assume the investor wants to maximize return and minimize risk after paying transaction costs. At the same time, he/she requires that the portfolio liquidity is not less than a given constant, after rebalancing of the existing portfolio. Based on the above, the portfolio rebalancing problem is formulated as follows:

(P1) $\max \sum_{i=1}^{n} r_i(x_i^0 + x_i^+ - x_i^-) - \sum_{i=1}^{n} p(x_i^+ + x_i^-)$

$\min \sum_{t=1}^{T} \dfrac{|\sum_{i=1}^{n}(r_{it}-r_i)x_i| + \sum_{i=1}^{n}(r_i-r_{it})x_i}{2T}$

s.t. $\sum_{j=1}^{n}(\frac{la_j+lb_j}{2} + \frac{\beta_j-\alpha_j}{6})x_j \geq l,$

$\sum_{i=1}^{n}(x_i^0 + x_i^+ - x_i^-) + \sum_{i=1}^{n} p(x_i^+ + x_i^-) = 1,$

$x_i = x_i^0 + x_i^+ - x_i^-, \quad i = 1, 2, \cdots, n,$

$0 \leq x_i^+ \leq u_i, \quad i = 1, 2, \cdots, n,$

$0 \leq x_i^- \leq x_i^0, \quad i = 1, 2, \cdots, n.$

where l is a constant given by the investor.

Eliminating the absolute function of the second objective function, the above problem can be transformed into the following problem:

$$(\text{P2}) \max \sum_{j=1}^{n} r_j(x_j^0 + x_j^+ - x_j^-) - \sum_{j=1}^{n} p(x_j^+ + x_j^-)$$

$$\min \frac{1}{T} \sum_{t=1}^{T} y_t$$

$$\text{s.t.} \sum_{j=1}^{n} \left(\frac{la_j + lb_j}{2} + \frac{\beta_j - \alpha_j}{6}\right)x_j \geq l,$$

$$y_t + \sum_{i=1}^{n} (r_{it} - r_i)x_i \geq 0, \quad t = 1, 2, \cdots, T,$$

$$\sum_{i=1}^{n} (x_i^0 + x_i^+ - x_i^-) + \sum_{i=1}^{n} p(x_i^+ + x_i^-) = 1,$$

$$x_i = x_i^0 + x_i^+ - x_i^-, \quad i = 1, 2, \cdots, n,$$

$$0 \leq x_i^+ \leq u_i, \quad i = 1, 2, \cdots, n,$$

$$0 \leq x_i^- \leq x_i^0, \quad i = 1, 2, \cdots, n,$$

$$y_t \geq 0, \quad t = 1, 2, \cdots, T.$$

where l is a constant given by the investor.

The above problem is a bi-objective linear programming problem. One can use several multiple objective linear programming algorithms to solve it efficiently.

6.3 Portfolio Rebalancing Model based on Fuzzy Decision

Since investment is generally influenced by changes in social and economic circumstances, an optimization approach is not always the best. In some cases, a satisfaction approach is much better than an optimization one. An investor always has aspiration levels for expected return and risk. In the real world of financial management, expert's knowledge and experience are very important in decision-making. Based on experts' knowledge, the investor may decide his/her aspiration levels for expected portfolio return and risk. Watada employed a logistic function, i.e., a non-linear S shape membership function, to express aspiration levels of an investor's expected return rate and risk. The S shape membership function is given by:

$$f(x) = \frac{1}{1 + \exp(-\alpha x)}. \tag{6.16}$$

The function has a shape similar to the tangent hyperbolic function employed by H. Leberling, but it is more easily handled than the tangent hyperbolic function. Therefore, it is more appropriate to consider the logistic function to denote a vague goal level, which an investor may consider. According to the maximization principle, and using variance to measure the portfolio risk, Watada proposed a fuzzy portfolio selection model. The model extended Markowitz's mean-variance model to the fuzzy case.

In the portfolio rebalancing model proposed in Section 3, the two objectives (return and risk) and the constraint on the liquidity of the portfolio are considered. Since the expected return, the risk and the liquidity are vague and uncertain, we use the non-linear S shape membership functions proposed by Watada to express the aspiration levels of expected return, risk and liquidity of the portfolio. Using the semi-absolute deviation risk function to measure the portfolio risk, we propose a fuzzy portfolio rebalancing model based on Bellman-Zadeh's maximization principle.

The membership functions of the goals for expected return, risk and liquidity are given as follows.

- a) Membership function of the goal for expected portfolio return

$$\mu_r(x) = \frac{1}{1 + \exp(-\alpha_r(E(r(x)) - r_M))}. \tag{6.17}$$

where r_M is the mid-point where the membership function value is 0.5 and α_r can be given by the investor based on his/her own degree of satisfaction for the expected return. r_M represents the middle aspiration level for the portfolio return. Fig 1 shows the membership function of the goal for expected return.

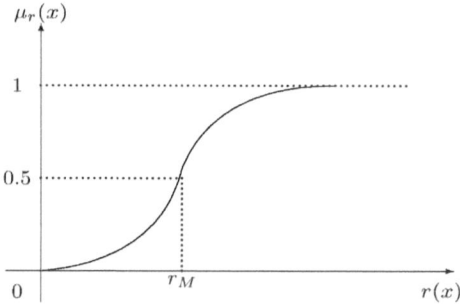

Fig. 6.1. Membership function of the goal for expected return

- b) Membership function of the goal for portfolio risk

$$\mu_w(x) = \frac{1}{1 + \exp(\alpha_w(w(x) - w_M))}. \tag{6.18}$$

where w_M is the mid-point where the membership function value is 0.5 and α_w can be given by the investor based on his/her own degree of satisfaction regarding the level of risk. w_M represents the middle aspiration level for portfolio risk. Fig 2 shows the membership function of the goal for risk.
- c) Membership function for the goal for portfolio liquidity

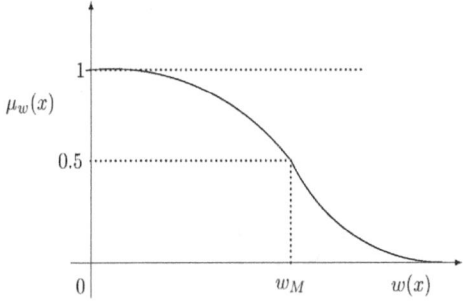

Fig. 6.2. Membership function of the goal for risk

$$\mu_{\hat{i}}(x) = \frac{1}{1 + \exp(-\alpha_l(E(\hat{l}(x)) - l_M))}.$$ (6.19)

where l_M is the mid point where the membership function value is 0.5 and α_l can be given by the investor based on his/her own degree of satisfaction regarding liquidity. l_M represents the middle aspiration level for the portfolio liquidity. Fig 3 shows the membership function of the goal for liquidity.

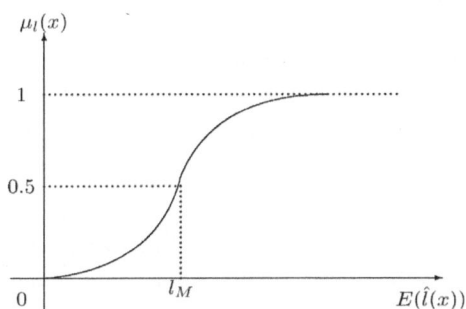

Fig. 6.3. Membership function for the goal for liquidity

Remark: α_r, α_w and α_l determine the shapes of membership functions $\mu_r(x), \mu_w(x)$ and $\mu_{\hat{i}}(x)$ respectively, where $\alpha_r > 0, \alpha_w > 0$ and $\alpha_l > 0$. As parameters α_r, α_w and α_l become larger, their vagueness declines.

According to Bellman and Zadeh's maximization principle, we can define

$$\eta = \min\{\mu_r(x), \mu_w(x), \mu_{\hat{i}}(x)\}.$$ (6.20)

The fuzzy portfolio rebalancing problem can be formulated as follows:

(P3) max η

s.t. $\mu_r(x) \geq \eta$,

$\mu_w(x) \geq \eta$,

$\mu_{\hat{l}}(x) \geq \eta$,

$\sum_{i=1}^{n}(x_i^0 + x_i^+ - x_i^-) + \sum_{i=1}^{n} p(x_i^+ + x_i^-) = 1$,

$x_i = x_i^0 + x_i^+ - x_i^-$, $i = 1, 2, \cdots, n$,

$0 \leq x_i^+ \leq u_i$, $i = 1, 2, \cdots, n$,

$0 \leq x_i^- \leq x_i^0$, $i = 1, 2, \cdots, n$,

$0 \leq \eta \leq 1$.

By (6.17), (6.18) and (6.19), the fuzzy portfolio rebalancing problem can be rewritten as follows:

(P4) max η

s.t. $\eta + \exp(-\alpha_r(E(r(x)) - r_M))\eta \leq 1$,

$\eta + \exp(\alpha_w(w(x) - w_M))\eta \leq 1$,

$\eta + \exp(-\alpha_l(E(\hat{l}(x)) - l_M))\eta \leq 1$,

$\sum_{i=1}^{n}(x_i^0 + x_i^+ - x_i^-) + \sum_{i=1}^{n} p(x_i^+ + x_i^-) = 1$,

$x_i = x_i^0 + x_i^+ - x_i^-$, $i = 1, 2, \cdots, n$,

$0 \leq x_i^+ \leq u_i$, $i = 1, 2, \cdots, n$,

$0 \leq x_i^- \leq x_i^0$, $i = 1, 2, \cdots, n$,

$0 \leq \eta \leq 1$.

where α_r, α_w and α_l are parameters which can be given by the investor based on his/her own degree of satisfaction regarding the three factors.

Substituting $\theta = \log\frac{\eta}{1-\eta}$, then $\eta = \frac{1}{1+\exp(-\theta)}$. The S shape membership function is monotonously increasing, so maximizing η maximizes θ. Therefore, the above problem may be transformed to an equivalent problem, as follows:

(P5) max θ

s.t. $\alpha_r(\sum\limits_{i=1}^{n} r_i x_i - \sum\limits_{i=1}^{n} p(x_i^+ + x_i^-)) - \theta \geq \alpha_r r_M,$

$\theta + \frac{\alpha_w}{T} \sum\limits_{t=1}^{T} y_t \leq \alpha_w w_M,$

$\alpha_l \sum\limits_{j=1}^{n} (\frac{la_j + lb_j}{2} + \frac{\beta_j - \alpha_j}{6}) x_j - \theta \geq \alpha_l l_M,$

$y_t + \sum\limits_{i=1}^{n} (r_{it} - r_i) x_i \geq 0, \quad t = 1, 2, \cdots, T,$

$\sum\limits_{i=1}^{n} (x_i^0 + x_i^+ - x_i^-) + \sum\limits_{i=1}^{n} p(x_i^+ + x_i^-) = 1,$

$x_i = x_i^0 + x_i^+ - x_i^-, \quad i = 1, 2, \cdots, n,$

$0 \leq x_i^+ \leq u_i, \quad i = 1, 2, \cdots, n,$

$0 \leq x_i^- \leq x_i^0, \quad i = 1, 2, \cdots, n,$

$y_t \geq 0, \quad t = 1, 2, \cdots, T,$

$\theta \geq 0.$

where α_r, α_w and α_l are parameters which can be given by the investor based on his/her own degree of satisfaction regarding the three factors.

Problem (P5) is also a standard linear programming problem. One can use several linear programming algorithms to solve it efficiently; for example, the simplex method.

Remark: The non-linear S shape membership functions of the three factors may change shape according to parameters α_r, α_w and α_l. By selecting the values of these parameters, the aspiration levels of the three factors can be described accurately. On the other hand, different parameter values may reflect investors' different aspiration levels. Therefore, the proposed portfolio rebalancing model is convenient for different investors to formulate their individual investment strategies.

6.4 Numerical Example

In this section, we give a numerical example to illustrate the proposed portfolio rebalancing model. Assume that an investor chooses 30 different stocks from the Shanghai Stock Exchange for his/her investment. The exchange codes of the 30 stocks are given in Table 1.

The rate of transaction costs for stocks is 0.0055 in the two securities markets in the Chinese mainland. Assume that the investor already owns an existing portfolio and he/she will not invest additional capital in the portfolio rebalancing process. The exchange codes of the stocks in the existing portfolio and the proportions of the stocks are listed in Table 2.

The financial market situation changes, meaning that the investor needs to change his/her investment strategy. In our example, we assume that the upper bound of the proportion of Stock j owned by the investor is 1. Now

Table 6.1. Exchange codes of 30 stocks

Stock	1	2	3	4	5	6	7	8
Code	600058	600061	600062	600070	600071	600073	600079	600081
Stock	9	10	11	12	13	14	15	16
Code	600085	600091	600094	600098	6000100	600619	600621	600624
Stock	17	18	19	20	21	22	23	24
Code	600630	600636	600637	600827	600831	600834	600843	600847
Stock	25	26	27	28	29	30		
Code	600850	600853	600857	600860	600864	600874		

Table 6.2. The exchange codes and the proportions of stocks in the existing portfolio

Code	600058	600061	600062	600070	600071	600073	600079
Proportions	0.05	0.08	0.05	0.35	0.1	0.12	0.25

we use the fuzzy portfolio rebalancing model proposed in this study to restructure the investor's assets. First, we collect historical data on the thirty stocks from January, 1999 to January, 2002. The data are downloaded from the web-site www.stockstar.com. Then we use one month as a period to obtain the historical rates of returns for thirty-six periods. The expected rates of returns based on these historical data are listed in Table 3.

Table 6.3. The expected return rates of the stocks

Stock	1	2	3	4	5	6	7	8	9	10
r	0.0376	0.0314	0.0326	0.0226	0.0285	0.0495	0.0220	0.0478	0.0271	0.0344
Stock	11	12	13	14	15	16	17	18	19	20
r	0.0225	0.0293	0.0411	0.0214	0.0225	0.0285	0.0241	0.0274	0.0295	0.0318
Stock	21	22	23	24	25	26	27	28	29	30
r	0.0463	0.0391	0.0244	0.0326	0.0366	0.0229	0.0361	0.0274	0.0312	0.0321

Since we assume that the future turnover rates of the securities are trapezoidal fuzzy numbers, we need to estimate the tolerance interval, left width and right width of the fuzzy numbers. In the real world of portfolio management, the investor can obtain values of these parameters by using the Delphi Method based on experts' knowledge. In our example, based on historical data of the securities' turnover rates, we adopt the frequency statistical method to estimate these parameters. In the following, we give the estimation method for the fuzzy turnover rates for Stock 1 in detail. First, we use historical data (daily turnover rates from January, 1999 to January, 2002) to calculate the frequency of historical turnover rates. Note that most of the historical turnover rates fall into the intervals $[0.008, 0.010]$, $[0.010, 0.012]$, $[0.012, 0.014]$, $[0.014, 0.016]$ and $[0.016, 0.018]$. We regard the midpoints of the intervals $[0.008, 0.010]$ and $[0.016, 0.018]$ as the left and the right endpoints of the tolerance interval, respectively. Therefore, the tolerance interval of the fuzzy turnover rate is $[0.009, 0.017]$. By observing all the historical data, we use 0.001 and 0.035 as the minimum and the maximum possible values of uncertain turnover rates in the future. Thus, the left width is 0.008 and the right width is 0.018. The fuzzy turnover rate of Stock 1 is $(0.009, 0.017, 0.008, 0.018)$. Using a similar method, we obtain the fuzzy turnover rates of all 30 stocks. These are listed in Table 4.

Table 6.4. The fuzzy turnover rates of the stocks

stock	1	2
$\hat{l}$	(0.009, 0.017, 0.008, 0.018)	(0.012,0.025, 0.01, 0.023)
Stock	3	4
$\hat{l}$	(0.009, 0.017, 0.0082, 0.017)	(0.010, 0.019, 0.008, 0.018)
Stock	5	6
$\hat{l}$	(0.013, 0.027, 0.010, 0.019)	(0.010, 0.025, 0.008, 0.018)
Stock	7	8
$\hat{l}_7$	(0.010, 0.020, 0.009, 0.019)	(0.008, 0.016, 0.007, 0.015)
Stock	9	10
$\hat{l}$	(0.005, 0.015, 0.0045, 0.017)	(0.007, 0.017, 0.006, 0.016)
Stock	11	12
$\hat{l}$	(0.011, 0.019, 0.009, 0.018)	(0.009, 0.024, 0.008, 0.025)
Stock	13	14
$\hat{l}$	(0.007, 0.016, 0.0064, 0.018)	(0.013, 0.032, 0.011, 0.030)
Stock	15	16
$\hat{l}$	(0.011, 0.026, 0.009, 0.028)	(0.012, 0.031, 0.011, 0.030)
Stock	17	18
$\hat{l}$	(0.006, 0.031, 0.005, 0.026)	(0.008, 0.015, 0.0076, 0.017)
Stock	19	20
$\hat{l}$	(0.011, 0.050, 0.0095, 0.047)	(0.008, 0.025, 0.006, 0.026)
Stock	21	22
$\hat{l}$	(0.010, 0.023, 0.008, 0.021)	(0.011, 0.031, 0.010, 0.030)
Stock	23	24
$\hat{l}$	(0.012, 0.046, 0.010, 0.043)	(0.009, 0.026, 0.008, 0.019)
Stock	25	26
$\hat{l}$	(0.007, 0.027, 0.0065, 0.025)	(0.010, 0.036, 0.009, 0.028)
Stock	27	28
$\hat{l}$	(0.011, 0.029, 0.010, 0.026)	(0.008, 0.043, 0.007, 0.035)
Stock	29	30
$\hat{l}$	(0.007, 0.035, 0.006, 0.034)	(0.006, 0.031, 0.0045, 0.029)

In the following, we present two kinds of computational results derived from conservative and aggressive techniques. First, we assume that the investor has a conservative and pessimistic mind. Then the values of r_M, w_M and l_M, given by the investor are small. They are as follows:

$$r_M = 0.029, w_M = 0.030, l_M = 0.0225.$$

Considering the three factors (return, risk and liquidity) as fuzzy numbers with a non-linear membership function, we obtain a portfolio rebalancing strategy by solving (P5). In the example, we give three different values of parameters α_r, α_w and α_l. The corresponding computational results are listed in Tables 5-8.

Table 6.5. Membership grade η, obtained risk, obtained expected return and obtained liquidity when $r_M = 0.029$, $w_M = 0.030$ and $l_M = 0.0225$

η	θ	α_r	α_w	α_l	obtained risk	obtained expected return	obtained liquidity
0.811	1.454	600	800	600	0.0282	0.0314	0.0304
0.806	1.425	500	1000	500	0.0286	0.0319	0.0303
0.785	1.295	400	1200	400	0.0289	0.0322	0.0302

Table 6.6. Portfolio rebalancing ratio when $r_M = 0.029$, $w_M = 0.030$, $l_M = 0.0225$, $\alpha_r = 600$, $\alpha_w = 800$ and $\alpha_l = 600$

Stock	1	2	3	4	5	6	7	8	9	10
Purchase ratio	0.000	0.000	0.092	0.000	0.000	0.000	0.000	0.294	0.000	0.021
Sell ratio	0.016	0.000	0.000	0.035	0.100	0.051	0.250	0.000	0.000	0.000
Stock	11	12	13	14	15	16	17	18	19	20
Purchase ratio	0.000	0.000	0.000	0.000	0.000	0.000	0.000	0.000	0.000	0.000
Sell ratio	0.000	0.000	0.000	0.000	0.000	0.000	0.000	0.000	0.000	0.000
Stock	21	22	23	24	25	26	27	28	29	30
Purchase ratio	0.100	0.044	0.000	0.000	0.000	0.000	0.000	0.000	0.038	0.177
Sell ratio	0.000	0.000	0.000	0.000	0.000	0.000	0.000	0.000	0.000	0.000

Next, we assume that the investor has an aggressive and optimistic mind. Then the values of r_M, w_M and l_M, given by the investor, are large. They are as follows:

$$r_M = 0.032, w_M = 0.034, l_M = 0.026.$$

Considering the three factors (return, risk and liquidity) as fuzzy numbers with a non-linear membership function, we obtain a portfolio rebalancing strategy by solving (P5). In the example, we give three different values of

Table 6.7. Portfolio rebalancing ratio when $r_M = 0.029$, $w_M = 0.030$, $l_M = 0.0225$, $\alpha_r = 500$, $\alpha_w = 1000$ and $\alpha_l = 500$.

Stock	1	2	3	4	5	6	7	8	9	10
Purchase ratio	0.000	0.000	0.086	0.000	0.000	0.000	0.000	0.291	0.000	0.009
Sell ratio	0.014	0.000	0.000	0.035	0.100	0.040	0.250	0.000	0.000	0.000
Stock	11	12	13	14	15	16	17	18	19	20
Purchase ratio	0.000	0.000	0.000	0.000	0.000	0.000	0.000	0.000	0.000	0.000
Sell ratio	0.000	0.000	0.000	0.000	0.000	0.000	0.000	0.000	0.000	0.000
Stock	21	22	23	24	25	26	27	28	29	30
Purchase ratio	0.110	0.050	0.000	0.000	0.000	0.000	0.000	0.000	0.038	0.171
Sell ratio	0.000	0.000	0.000	0.000	0.000	0.000	0.000	0.000	0.000	0.000

Table 6.8. Portfolio rebalancing ratio when $r_M = 0.029$, $w_M = 0.030$, $l_M = 0.0225$, $\alpha_r = 400$, $\alpha_w = 1200$ and $\alpha_l = 400$.

Stock	1	2	3	4	5	6	7	8	9	10
Purchase ratio	0.000	0.000	0.079	0.000	0.000	0.000	0.000	0.289	0.000	0.000
Sell ratio	0.013	0.000	0.000	0.035	0.100	0.030	0.250	0.000	0.000	0.000
Stock	11	12	13	14	15	16	17	18	19	20
Purchase ratio	0.000	0.000	0.000	0.000	0.000	0.000	0.000	0.000	0.000	0.000
Sell ratio	0.000	0.000	0.000	0.000	0.000	0.000	0.000	0.000	0.000	0.000
Stock	21	22	23	24	25	26	27	28	29	30
Purchase ratio	0.117	0.055	0.000	0.000	0.000	0.000	0.000	0.000	0.037	0.166
Sell ratio	0.000	0.000	0.000	0.000	0.000	0.000	0.000	0.000	0.000	0.000

parameters α_r, α_w and α_l. The corresponding computational results are listed in Tables 9-12.

Table 6.9. Membership grade η, obtained risk, obtained expected return and obtained liquidity when $r_M = 0.032$, $w_M = 0.034$ and $l_M = 0.026$.

η	θ	α_r	α_w	α_l	obtained risk	obtained expected return	obtained liquidity
0.849	1.726	600	800	600	0.0318	0.0349	0.0295
0.836	1.630	500	1000	500	0.0324	0.0353	0.0293
0.802	1.396	400	1200	400	0.0328	0.0355	0.0295

Since it is possible that the non-linear S shape membership function changes its shape according to the values of the parameters, the non-linear membership function can reflect the investor's mind accurately and suitably. From the above results, we find that we can obtain the different portfolio rebalancing strategies by solving (P5) in which the different values of parameters (α_r, α_w and α_l) are given. By choosing the values of the parameters α_r, α_w

Table 6.10. Portfolio rebalancing ratio when $r_M = 0.032$, $w_M = 0.034$, $l_M = 0.026$, $\alpha_r = 600$, $\alpha_w = 800$ and $\alpha_l = 600$.

Stock	1	2	3	4	5	6	7	8	9	10
Purchase ratio	0.000	0.000	0.000	0.000	0.000	0.000	0.000	0.311	0.000	0.000
Sell ratio	0.000	0.000	0.000	0.035	0.100	0.000	0.250	0.000	0.000	0.000
Stock	11	12	13	14	15	16	17	18	19	20
Purchase ratio	0.000	0.000	0.000	0.000	0.000	0.000	0.000	0.000	0.000	0.000
Sell ratio	0.000	0.000	0.000	0.000	0.000	0.000	0.000	0.000	0.000	0.000
Stock	21	22	23	24	25	26	27	28	29	30
Purchase ratio	0.185	0.098	0.000	0.000	0.000	0.000	0.000	0.000	0.026	0.081
Sell ratio	0.000	0.000	0.000	0.000	0.000	0.000	0.000	0.000	0.000	0.000

Table 6.11. Portfolio rebalancing ratio when $r_M = 0.032$, $w_M = 0.034$, $l_M = 0.026$, $\alpha_r = 500$, $\alpha_w = 1000$ and $\alpha_l = 500$.

Stock	1	2	3	4	5	6	7	8	9	10
Purchase ratio	0.000	0.000	0.000	0.000	0.000	0.000	0.000	0.324	0.000	0.000
Sell ratio	0.000	0.000	0.000	0.035	0.100	0.000	0.250	0.000	0.000	0.000
Stock	11	12	13	14	15	16	17	18	19	20
Purchase ratio	0.000	0.000	0.000	0.000	0.000	0.000	0.000	0.000	0.000	0.000
Sell ratio	0.000	0.000	0.000	0.000	0.000	0.000	0.000	0.000	0.000	0.000
Stock	21	22	23	24	25	26	27	28	29	30
Purchase ratio	0.196	0.099	0.000	0.000	0.000	0.000	0.000	0.000	0.020	0.061
Sell ratio	0.000	0.000	0.000	0.000	0.000	0.000	0.000	0.000	0.000	0.000

Table 6.12. Portfolio rebalancing ratio when $r_M = 0.032$, $w_M = 0.034$, $l_M = 0.026$, $\alpha_r = 400$, $\alpha_w = 1200$ and $\alpha_l = 400$.

Stock	1	2	3	4	5	6	7	8	9	10
Purchase ratio	0.000	0.000	0.000	0.000	0.000	0.000	0.000	0.344	0.000	0.000
Sell ratio	0.020	0.000	0.000	0.035	0.100	0.000	0.250	0.000	0.000	0.000
Stock	11	12	13	14	15	16	17	18	19	20
Purchase ratio	0.000	0.000	0.000	0.000	0.000	0.000	0.000	0.000	0.000	0.000
Sell ratio	0.000	0.000	0.000	0.000	0.000	0.000	0.000	0.000	0.000	0.000
Stock	21	22	23	24	25	26	27	28	29	30
Purchase ratio	0.214	0.101	0.000	0.000	0.000	0.000	0.000	0.000	0.024	0.038
Sell ratio	0.000	0.000	0.000	0.000	0.000	0.000	0.000	0.000	0.000	0.000

and α_l according to the investor's frame of mind, it is possible to achieve a favorite portfolio rebalancing strategy.

6.5 Conclusion

In addition to the more usual factors of expected return and risk, portfolio liquidity is considered in the portfolio rebalancing process. The turnover rates of securities are used to measure their liquidity. Considering all the three factors, a linear programming model for portfolio rebalancing with transaction costs is proposed. An investor's aspiration levels for the expected return, risk and liquidity are vague in an uncertain environment. The vague aspiration levels are considered to be fuzzy numbers with a non-linear S shape membership function. Furthermore, based on the fuzzy decision theory, a fuzzy portfolio rebalancing model with transaction costs is proposed. An example is given to illustrate the behavior of the proposed fuzzy portfolio rebalancing model using real data from the Shanghai Stock Exchange. The computation results show that the portfolio rebalancing model with a non-linear S shape membership function can generate a favorite portfolio rebalancing strategy according to the investor's degree of satisfaction.

7

Fuzzy Mixed Projects and Securities Portfolio Selection Model

7.1 Introduction

During the past decade, there has been a dramatic increase in institutional investments. Although most of those investments remain focused on traditional securities, there is an increase in various forms of alternative investment classes, e.g., venture capital, private equity, private debt and real estate, etc. With the diversification into different types of assets, the overall portfolio risk can be lowered while the potential for more benefits can be increased over the long term.

The mean variance methodology for portfolio selection has been central to research activities in the traditional securities investment field and has served as a basis for development of the modern financial theory over the past five decades. Konno used the absolute deviation risk function to replace the risk function in Markowitz's model to formulate a mean absolute deviation portfolio optimization model. In today's extremely competitive business environment, investors may consider investing their funds in other kinds of assets, besides securities. Byrne and Lee and Keng found that mixed assets portfolio, including listed property trusts, direct property and financial assets always dominated the financial assets portfolio. Selection of projects for portfolio selection is one of the most important decision problems which the corporations face. In recent years, some researchers studied project portfolio selection problems by using mathematical programming methods, e.g., Coffin and Taylor III, and Ringuest, Graves and Case, etc. Some securities and projects can be integrated into a mixed assets portfolio. Reyck, Degraeve and Gustafsson proposed a mixed assets portfolio selection model involving projects and securities. Transaction cost is one of the main sources of concern to portfolio managers. Arnott and Wagner found that ignoring transaction costs would result in an inefficient portfolio. Yoshimoto's empirical analysis also drew the same conclusion. However, transaction costs are not considered in Reyck, Degraeve and Gustafsson's mixed asset portfolio selection model.

In this chapter, considering the proportional transaction costs, we will use the expected return and the semi-absolute deviation risk as objective functions and shall propose a bi-objective programming model for the mixed assets portfolio selection problem. Furthermore, we use fuzzy numbers to describe investors' vague aspiration levels for the expected return and the semi-absolute deviation risk and propose a fuzzy mixed assets portfolio selection model.

7.2 Bi-objective Programming Model for Mixed Asset Portfolio Selection

We assume that an investor allocates his/her wealth among traditional securities and projects. Hence, in the mixed assets portfolio selection problem, assets available for investment are divided into two types of assets. The first class of assets consists of traditional securities. The second class of assets consists of projects. The main difference between these two types of investments is that the decision variables for projects are binary, while those for securities are continuous. There is a capital budget for each project before investment. The capital budgets can be given by investors or some experts. We assume that the cost of carrying out a project will be the corresponding capital budget once the project is started. That is, once the investor decides to invest in a project, the amount earmarked for the project must be the capital budget of that project. In addition, investment in these projects cannot be reallocated at any time, while investments in securities can be reallocated at any time.

We assume that the securities component of the mixed asset is composed of n risky securities $S_i, i = 1, 2, \cdots, n$, offering random rates of returns and a risk-free security S_{n+1} offering a fixed rate of return. The projects component is composed of m projects $P_j, j = 1, 2, \cdots, m$. Assume the investor starts with an existing portfolio, which only includes securities, and then decides how to reconstruct a new mixed asset portfolio with securities and projects. We introduce some notations as follows.

$\widetilde{r_i}$: the random variable representing the rate of return on the security $S_i, i = 1, 2, \cdots, n$ without transaction costs;

r_{n+1}: the rate of return of the risk-free security S_{n+1};

r_i: the rate of expected return on the security $S_i, i = 1, 2, \cdots, n$ without transaction costs;

$\widetilde{R_j}$: the random variable representing the random net return on the project $P_j, j = 1, 2, \cdots, m$ after removing the cost (the budget);

R_j: the expected net return on the project $P_j, j = 1, 2, \cdots, m$ after removing the cost (the budget);

M: the total amount of assets owned by the investor;

X_i: the amount of the total investment devoted to the risky security $S_i, i = 1, 2, \cdots, n$ and the risk-free security S_{n+1};

x_i: the proportion of the total investment devoted to the risky security $S_i, i = 1, 2, \cdots, n$ and the risk-free security S_{n+1}, i.e., $x_i = \frac{X_i}{M}$;

X_i^0: the amount of the total investment devoted to the risky security $S_i, i = 1, 2, \cdots, n$ and the risk-free security S_{n+1} in the existing portfolio;

x_i^0: the proportion of the total investment devoted to the risky security $S_i, i = 1, 2, \cdots, n$ and the risk-free security S_{n+1} in the existing portfolio;

k_i: the rate of transaction costs for the risky security $S_i, i = 1, 2, \cdots, n$ and the risk-free security S_{n+1};

z_j: the binary variable indicating whether project $P_j, j = 1, 2, \cdots, m$ is started or not,

$$z_j = \begin{cases} 1 \text{ if project } P_j \text{ is selected for funding,} \\ 0 \text{ otherwise.} \end{cases}$$

We assume that the vector of random variables $(\widetilde{r}_1, \widetilde{r}_2, \cdots, \widetilde{r}_n, \widetilde{R}_1, \widetilde{R}_2, \cdots, \widetilde{R}_m)$ is distributed over the finite sample space $\{(r_{1t}, \cdots, r_{nt}, R_{1t}, \cdots, R_{mt}), t = 1, 2, \cdots, T\}$ and the probabilities

$$p_t = P_r\{(\widetilde{r}_1, \cdots, \widetilde{r}_n, \widetilde{R}_1, \cdots, \widetilde{R}_m) = (r_{1t}, \cdots, r_{nt}, R_{1t}, \cdots, R_{mt})\}, t = 1, 2, \cdots, T$$

are known. Then the expected rate of return r_i of the risky security $S_i, i = 1, 2, \cdots, n$ without transaction costs is given by

$$r_i = \sum_{t=1}^{T} p_t r_{it}, \ i = 1, 2, \cdots, n,$$

where r_{it} can be determined by forecast data. The expected net return R_j on the project $P_j, j = 1, 2, \cdots, m$ is given by

$$R_j = \sum_{t=1}^{T} p_t R_{jt}, \ j = 1, 2, \cdots, m,$$

where R_{jt} can be determined by forecast data.

Given a mixed asset portfolio $(x_1, x_2, \cdots, x_n, x_{n+1}, z_1, z_2, \cdots, z_m)$, the expected return of the portfolio without transaction costs can be expressed by

$$\sum_{i=1}^{n+1} r_i X_i + \sum_{j=1}^{m} R_j z_j = \sum_{i=1}^{n+1} \sum_{t=1}^{T} p_t r_{it} X_i + \sum_{j=1}^{m} \sum_{t=1}^{T} p_t R_{jt} z_j,$$

where $X_i = Mx_i, \ i = 1, 2, \cdots, n+1$ and $r_{n+1,t} = r_{n+1}, \ t = 1, 2, \cdots, T$.

We use a V shape function to express the transaction costs. Specifically we let the transaction costs of the security $S_i, i = 1, 2, \cdots, n, n+1$ be given by

$$C_i(X_i) = k_i |X_i - X_i^0|.$$

Hence the total transaction costs of the mixed asset portfolio are expressed as

$$\sum_{i=1}^{n+1} C_i(X_i) = \sum_{i=1}^{n+1} k_i |X_i - X_i^0|.$$

Let $x = (x_1, \cdots, x_{n+1})$, $z = (z_1, \cdots, z_m)$ and $X = (X_1, \cdots, X_{n+1})$. Then the expected net return on the mixed asset portfolio after paying the transaction costs is given by

$$f(X, z) = \sum_{i=1}^{n+1} r_i X_i + \sum_{j=1}^{m} R_j z_j - \sum_{i=1}^{n+1} C_i(X_i)$$

$$= \sum_{i=1}^{n+1} \sum_{t=1}^{T} p_t r_{it} X_i + \sum_{j=1}^{m} \sum_{t=1}^{T} p_t R_{jt} z_j - \sum_{i=1}^{n+1} k_i |X_i - X_i^0|.$$

If we use $x_1, \cdots, x_n, x_{n+1}, x_1^0, \cdots, x_n^0, x_{n+1}^0$ instead of $X_1, \cdots, X_n, X_{n+1}, X_1^0, \cdots, X_n^0, X_{n+1}^0$, respectively, then the expected net return on the mixed asset portfolio after paying the transaction costs, is also given by

$$f(x, z) = \sum_{i=1}^{n+1} \sum_{t=1}^{T} p_t r_{it} x_i M + \sum_{j=1}^{m} \sum_{t=1}^{T} p_t R_{jt} z_j - \sum_{i=1}^{n+1} k_i M |x_i - x_i^0|.$$

Maximizing the expected net return $f(x, z)$ on the mixed asset portfolio after paying the transaction costs can be considered an objective of the mixed asset portfolio selection problem.

In the traditional securities portfolio selection, Markowitz used variance to measure the risk of a portfolio, which is the first quantitative measure of a risk. Subsequently, several other risk measures have been proposed in the literature on financial portfolio selection. These methods include semi-variance, absolute deviation, semi-absolute deviation and so on. Since the semi-absolute deviation seems more suitable to measure the risk of a portfolio in practice, we will use the semi-deviation risk function in our mixed asset portfolio selection model.

The semi-absolute deviation of return on the mixed asset portfolio below the expected return at state t, $t = 1, 2, \cdots, T$ can be represented as

$$W_t(X, z) = \left| \min\{0, \sum_{i=1}^{n} (r_{ti} - r_i) X_i + \sum_{j=1}^{m} (R_{tj} - R_j) z_j\} \right|.$$

So the expected semi-absolute deviation of the return on the mixed asset portfolio below the expected return can be represented as

$$W(X, z) = \sum_{t=1}^{T} p_t W_t(X, z)$$

$$= \sum_{t=1}^{T} p_t \left| \min\{0, \sum_{i=1}^{n} (r_{ti} - r_i) X_i + \sum_{j=1}^{m} (R_{tj} - R_j) z_j\} \right|.$$

Let $w(x, z) = \frac{W(X, z)}{M}$. Then we have

$$w(x, z) = \sum_{t=1}^{T} p_t \left| \min\{0, \sum_{i=1}^{n} (r_{ti} - r_i) x_i + \sum_{j=1}^{m} z_j \frac{R_{tj} - R_j}{M}\} \right|.$$

In this paper, we adopt the function $w(x, z)$ to measure the risk of the mixed asset portfolio. Minimizing the risk of the mixed asset portfolio can be considered another objective of the mixed asset portfolio selection problem.

In the mixed asset portfolio selection problem with securities and projects, we consider the following constraints. First, we introduce some notations.

B_j: the capital budget of project $P_j, j = 1, 2, \cdots, m$, i.e., the cost that the investor pays once the project is decided and started;

B: the maximum amount of investment devoted to the projects component in the mixed asset portfolio;

Y_j: the amount of the total investment devoted to project P_j, i.e.,

$$Y_j = B_j z_j, \quad j = 1, 2, \cdots, m;$$

S: the maximum amount of investment devoted to the securities component in the mixed asset portfolio;

• Capital budget constraint on projects component:

$$\sum_{j=1}^{m} Y_j = \sum_{j=1}^{m} B_j z_j \le B.$$

• Capital constraint on securities component:

$$\sum_{i=1}^{n+1} x_i M \le S.$$

• Total capital constraint:

$$\sum_{j=1}^{m} Y_j + \sum_{i=1}^{n+1} x_i M = \sum_{j=1}^{m} B_j z_j + \sum_{i=1}^{n+1} x_i M \le M.$$

• No short selling of securities:

$$x_i \ge 0, \quad i = 1, 2, \cdots, n+1.$$

Using the objectives and the constraints introduced in the previous subsection, the mixed asset portfolio selection problem can be formally stated as follows:

(BOP) $\max f(x,z) = \sum\limits_{i=1}^{n+1}\sum\limits_{t=1}^{T} p_t r_{it} x_i M + \sum\limits_{j=1}^{m}\sum\limits_{t=1}^{T} p_t R_{jt} z_j - \sum\limits_{i=1}^{n+1} k_i M |x_i - x_i^0|$

$\min\ w(x,z) = \sum\limits_{t=1}^{T} p_t |\min\{0, \sum\limits_{i=1}^{n}(r_{ti} - r_i)x_i + \sum\limits_{j=1}^{m} z_j \frac{R_{tj}-R_j}{M}\}|$

s.t. $\sum\limits_{j=1}^{m} B_j z_j \le B,$

$\sum\limits_{i=1}^{n+1} x_i M \le S,$

$\sum\limits_{j=1}^{m} B_j z_j + \sum\limits_{i=1}^{n+1} x_i M \le M,$

$x_i \ge 0,\ \ i = 1,2,\cdots,n+1,$

$z_j = \{0,1\},\ \ j = 1,2,\cdots,m.$

This problem is a bi-objective mixed-integer nonlinear programming problem.

The problem (BOP) can be reformulated as a bi-objective mixed-integer linear programming problem by using the following technique. Note that

$$|\min\{0, \sum\limits_{i=1}^{n}(r_{ti} - r_i)x_i + \sum\limits_{j=1}^{m} z_j \frac{R_{tj}-R_j}{M}\}|$$

$$= |\sum\limits_{i=1}^{n} \frac{(r_{ti}-r_i)x_i}{2} + \sum\limits_{j=1}^{m} \frac{z_j(R_{tj}-R_j)}{2M}| - \sum\limits_{i=1}^{n} \frac{(r_{ti}-r_i)x_i}{2} - \sum\limits_{j=1}^{m} \frac{z_j(R_{tj}-R_j)}{2M}.$$

Then, by introducing auxiliary variables $a_i^+, a_i^-, i = 1,2,\cdots,n+1$, and $xi_t^+, xi_t^-, t = 1,2,\cdots,T$, such that

$$a_i^+ + a_i^- = |x_i - x_i^0|,$$

$$a_i^+ - a_i^- = x_i - x_i^0,$$

$$a_i^+ \ge 0,\ \ a_i^- \ge 0,\ \ i = 1,2,\cdots,n+1,$$

$$xi_t^+ + xi_t^- = |\sum\limits_{i=1}^{n} \frac{(r_{ti}-r_i)x_i}{2} + \sum\limits_{j=1}^{m} \frac{z_j(R_{tj}-R_j)}{2M}|,$$

$$xi_t^+ - xi_t^- = \sum\limits_{i=1}^{n} \frac{(r_{ti}-r_i)x_i}{2} + \sum\limits_{j=1}^{m} \frac{z_j(R_{tj}-R_j)}{2M},$$

$$xi_t^+ \ge 0,\ \ xi_t^- \ge 0,\ \ t = 1,2,\cdots,T,$$

we may consider the following bi-objective mixed-integer linear programming problem:

$$(\text{BILP}) \ \max \ f(x,z) = \sum_{i=1}^{n+1}\sum_{t=1}^{T} p_t r_{it} x_i M + \sum_{j=1}^{m}\sum_{t=1}^{T} p_t R_{jt} z_j - \sum_{i=1}^{n+1} k_i M (a_i^+ + a_i^-)$$

$$\min \ w(x,z) = \sum_{t=1}^{T} 2 p_t x i_t^-$$

$$\text{s.t.} \ \ a_i^+ - a_i^- = x_i - x_i^0, \quad i = 1,2,\cdots,n+1,$$

$$x i_t^+ - x i_t^- = \sum_{i=1}^{n} \frac{(r_{ti} - r_i) x_i}{2} + \sum_{j=1}^{m} \frac{z_j (R_{tj} - R_j)}{2M}, \quad t = 1,2,\cdots,T,$$

$$\sum_{j=1}^{m} B_j z_j \leq B,$$

$$\sum_{i=1}^{n+1} x_i M \leq S,$$

$$\sum_{j=1}^{m} B_j z_j + \sum_{i=1}^{n+1} x_i M \leq M,$$

$$x_i \geq 0, \quad i = 1,2,\cdots,n+1,$$

$$a_i^+ \geq 0, \quad a_i^- \geq 0, \quad i = 1,2,\cdots,n+1,$$

$$x i_t^+ \geq 0, \quad x i_t^- \geq 0, \quad t = 1,2,\cdots,T,$$

$$z_j = \{0,1\}, \quad j = 1,2,\cdots,m.$$

It is not difficult to see that (BILP) is equivalent to (BOP). Thus the investor may determine his/her investment strategies by computing efficient solutions of (BILP).

7.3 Fuzzy Mixed Asset Portfolio Selection Model

In an investment, knowledge and experience of experts are very important in an investor's decision-making. Based on experts' knowledge, the investor may decide his/her levels of aspirations for the expected return and risk of a mixed asset portfolio. Watada employed a non-linear S shape membership function, to express aspiration levels of return and risk, which the investor would expect, and proposed a fuzzy active portfolio selection model. The S shape membership function is given by:

$$f(x) = \frac{1}{1 + \exp(-\alpha x)}.$$

In the bi-objective programming model of mixed asset portfolio selection proposed in Section 2, the two objectives, the expected return and the risk, are considered. Since the expected return and the risk are vague and uncertain, we use the non-linear S shape membership functions proposed by Watada to express the aspiration levels of the expected return and the risk.

The membership function of the expected return is given by

$$\mu_f(x,z) = \frac{1}{1 + \exp\left(-\alpha_f \left(f(x,z) - f_M\right)\right)},$$

where f_M is the mid-point where the membership function value is 0.5 and α_f can be given by the investor based on his/her own degree of satisfaction for the expected return.

The membership function of the risk is given by

$$\mu_w(x, z) = \frac{1}{1 + \exp(\alpha_w(w(x, z) - w_M))},$$

where w_M is the mid-point where the membership function value is 0.5 and α_w can be given by the investor based on his/her own degree of satisfaction regarding the level of risk.

Remark1: α_f and α_w determine the shapes of membership functions $\mu_f(x, z)$ and $\mu_w(x, z)$ respectively, where $\alpha_f > 0$ and $\alpha_w > 0$. The larger the parameters α_f and α_w, the lower is their vagueness.

According to Bellman and Zadeh's maximization principle, we can define

$$\lambda = \min\left\{\mu_f(x, z), \mu_w(x, z)\right\}.$$

The fuzzy mixed asset portfolio selection problem can be formulated as follows:

$$\text{(FP)} \quad \max \lambda$$
$$\text{s.t.} \quad \mu_f(x, z) \geq \lambda,$$
$$\mu_w(x, z) \geq \lambda,$$
$$\text{and all constraints of (BILP).}$$

Let $\eta = \log\frac{1}{1-\lambda}$, then $\lambda = \frac{1}{1+\exp(-\eta)}$. The logistic function is monotonously increasing, so maximizing λ makes η maximize. Therefore, the above problem can be transformed to an equivalent problem as follows:

$$\text{(FLP)} \quad \max \eta$$
$$\text{s.t.} \quad \alpha_f\left(f(x, z) - f_M\right) - \eta \geq 0,$$
$$\alpha_w\left(w(x, z) - w_M\right) + \eta \leq 0,$$
$$\text{and all constraints of (BILP),}$$

where α_f and α_w are parameters which can be given by the investor based on his/her own degree of satisfaction regarding the expected return and the risk.

(FLP) is a standard linear programming problem. One can use one of several algorithms of linear programming to solve it efficiently, for example, the simplex method.

Remark2: The non-linear S shape membership functions of the two factors may change their shape according to parameters α_f and α_w. Through selecting the values of these parameters, the aspiration levels of the two factors may be described accurately. On the other hand, different parameter values may reflect different levels of investors' aspiration. Therefore, it is convenient for different investors to formulate investment strategies by using the proposed fuzzy mixed asset portfolio selection model.

7.4 Numerical Example

Assume that the total value of assets M owned by an investor is \$300000. Originally, all the assets were invested in the risk-free security S_6. The rate r_6 of return of a risk-free security is 1.5%. Generally, the return of the risk-free security is less than the return of risky securities and projects. To obtain more profits, the investor reallocates his/her wealth among five risky securities $S_1, \cdots, S_5$, five projects $P_1, \cdots, P_5$ and the risk-free security S_6, where the rate $k_i, i = 1, 2, \cdots, 6$ of transaction costs is 0.4% for all securities.

Assume that there are eight possible scenarios of business environment in the future. Possible rates of returns on the five risky securities in these scenarios and the corresponding probabilities are listed in Table 1. Possible net returns on the five projects in these scenarios and the corresponding probabilities are listed in Table 2. The capital budgets of projects are listed in Table 3. The investor may stipulate the values of the maximum amounts of the investment, B and S, devoted to projects and securities components, according to his/her investment preference. In the following, we examine a case.

Table 7.1. Possible rates of returns and the expected rates of returns on the securities

State t	p_t	r_{t1}	r_{t2}	r_{t3}	r_{t4}	r_{t5}
1	0.100	-0.089	-0.007	-0.020	-0.011	-0.022
2	0.120	-0.042	0.043	0.036	-0.117	-0.053
3	0.120	0.120	0.047	0.128	-0.054	0.008
4	0.125	-0.062	-0.126	-0.090	0.109	0.057
5	0.125	0.147	0.230	-0.018	0.368	0.124
6	0.130	0.210	0.640	0.271	-0.135	0.277
7	0.130	0.011	-0.053	0.047	0.060	-0.105
8	0.150	0.005	-0.051	-0.017	0.014	-0.060
Expected Return (r_i)		0.041	0.092	0.043	0.030	0.028

Suppose that the maximum amount S of investment allocated to securities component is \$240000 and the maximum amount B of investment allocated to projects component is \$150000. We give the values of f_M and w_M, i.e., $f_M = 0.04, w_M = 0.0375$. In the example, we give three different values of parameters α_f and α_w. Using the above data, we computed satisfactory investment strategies by solving (FLP). All computations were carried out on a WINDOWS PC, using the LINDO solver. The membership grade, the obtained risk and the obtained return are listed in Table 7.4. The detailed allocations of the three optimal portfolios are shown in Table 7.5.

Table 7.2. Possible net returns and the expected net returns on the projects

State t	p_t	R_{t1}	R_{t2}	R_{t3}	R_{t4}	R_{t5}
1	0.100	$20000	-$20000	$2000	-$10000	$20000
2	0.120	$0	$10000	-$10000	$20000	-$20000
3	0.120	$40000	-$15000	$10000	-$20000	$10000
4	0.125	-$10000	$35000	$40000	$40000	$30000
5	0.125	-$5000	$40000	-$15000	$60000	$40000
6	0.130	-$15000	-$18000	$50000	-$15000	-$10000
7	0.130	$15000	$20000	-$8000	$10000	-$40000
8	0.150	$10000	$8000	$16000	-$30000	-$25000
Expected Return (R_j)		$6425	$8235	$10625	$10000	$11875

Table 7.3. The capital budgets of projects

B_1	B_2	B_3	B_4	B_5
$40000	$50000	$80000	$100000	$120000

Table 7.4. Membership grade λ, obtained risk and obtained expected return

λ	η	α_f	α_w	obtained risk	obtained expected return
0.999	14.0529	600	800	0.0146	0.0629
1.000	16.0195	500	1000	0.0215	0.0720
1.000	16.2212	400	1200	0.0240	0.0790

Since it is possible that the non-linear S shape membership function changes its shape according to the values of the parameters, the non-linear membership function can reflect the investor's mind accurately and suitably. From the above results, we can find that we get the different investment strategies by solving (FLP) in which the different values of the parameters α_f and α_w are given. Through choosing the values of the parameters α_r and α_w according to the investor's frame of mind, the investor may get a favorite investment strategy.

7.5 Conclusion

In today's extremely competitive business environment, investors have already invested in various classes of assets to keep their competitive advantages. Some

Table 7.5. The allocations of three optimal portfolios and the corresponding returns

Portfolio	x_1	x_2	x_3	x_4	x_5	x_6
Portfolio 1	0.0000	0.0316	0.2880	0.0000	0.3363	0.0440
Portfolio 2	0.0000	0.2343	0.0505	0.0000	0.3105	0.1048
Portfolio 3	0.0000	0.3862	0.0000	0.0000	0.1743	0.1395
Portfolio	z_1	z_2	z_3	z_4	z_5	
Portfolio 1	1	1	0	0	0	
Portfolio 2	1	1	0	0	0	
Portfolio 3	1	1	0	0	0	

securities and projects can be integrated into a mixed assets portfolio. The mixed asset portfolio increases the investors' benefit opportunities. Regarding the expected return and the risk, the two objective functions, we have proposed a bi-objective programming model for the mixed assets portfolio selection problem with transaction costs. Furthermore, investors' vague aspiration levels for the return and the risk are considered as fuzzy numbers. Based on the fuzzy decision theory, we have proposed a fuzzy mixed projects and securities portfolio selection model. A numerical example is given to illustrate the proposed fuzzy mixed assets portfolio selection model. The computation results show that the proposed model can generate a favorite mixed assets portfolio strategy, according to the investor's satisfaction degree.

Part III

Portfolio Selection Models with Interval
Coefficients

8

Linear Programming Model with Interval Coefficients

8.1 Introduction

Traditionally, it has been assumed that the distribution functions of possibility returns are known, while solving portfolio selection models. However, new securities and classes of assets have emerged in recent times and now, it is not always possible for an investor to specify them. In some cases, for instance, historical data of stocks are not available. In such cases, the uncertain returns of assets may be determined as interval numbers by using experts' knowledge. In recent years, some approaches for solving interval linear programming problems have been proposed by Ishibuchi and Tanaka, Tong, Liu.

In this chapter, we propose an interval semi-absolute deviation model for portfolio selection, where the expected returns of securities are treated as interval numbers. By introducing the concepts of pessimistic satisfaction index and optimistic satisfaction index of interval inequality relation, we convert the interval semi-absolute deviation portfolio selection problem into two parametric linear programming problems.

This chapter is organized as follows. In Section 2, we give some notations for interval numbers and briefly introduce some interval arithmetics. An order of relations over intervals is introduced. The concepts of pessimistic and optimistic satisfaction indices of interval inequality relations are given. Based on these concepts, an approach to compare interval numbers is proposed. In Section 3, an approach is presented for estimating the intervals of rates of returns of securities. In Section 4, an interval semi-absolute deviation model for portfolio selection is proposed. According to the approach proposed in Section 2, which concerns about comparing interval numbers, the interval portfolio selection problem is converted into two parametric linear programming problems. In Section 5, an example is given to illustrate our approach by using real data from the Shanghai Stock Exchange. A few concluding remarks are finally given in Section 6.

8.2 Notations and Definitions

Denote the set of all the real numbers by R. An order pair in a bracket defines an interval

$$a = [\underline{a}, \overline{a}] = \{x : \underline{a} \leq x \leq \overline{a}, x \in R\} \tag{8.1}$$

where $\underline{a}$ is the lower bound and $\overline{a}$ is the upper bound of interval a. The center and the width of a can be easily calculated as

$$m(a) = \frac{1}{2}(\underline{a} + \overline{a}) \text{ and } w(a) = \frac{1}{2}(\overline{a} - \underline{a}). \tag{8.2}$$

a can also be denoted by its center and width as

$$a = \langle m(a), w(a) \rangle = \{x : m(a) - w(a) \leq x \leq m(a) + w(a), x \in R\}. \tag{8.3}$$

Extension of ordinary arithmetic to closed intervals is known as interval arithmetic. For a detailed discussion, one can refer to Alefeld and Hansen. First, we quote a basic concept, as follows.

Definition 8.1(Alefeld): Let $\circ \in \{+, -, \times, \div\}$ be a binary operation on R. If a and b are two closed intervals, then

$$a \circ b = \{x \circ y : x \in a, y \in b\} \tag{8.4}$$

defines a binary operation on the set of all the closed intervals. In the case of division, it is always assumed that 0 is not in b.

The operations on intervals used in this paper are as follows:

$$a + b = [\underline{a} + \underline{b}, \overline{a} + \overline{b}], \tag{8.5}$$

$$a - b = [\underline{a} - \overline{b}, \overline{a} - \underline{b}], \tag{8.6}$$

$$a \pm k = [\underline{a} \pm k, \overline{a} \pm k], \tag{8.7}$$

$$ka = k[\underline{a}, \overline{a}] = \begin{cases} [k\underline{a}, k\overline{a}] & \text{for } k \geq 0 \\ [k\overline{a}, k\underline{a}] & \text{for } k < 0, \end{cases} \tag{8.8}$$

where k is a real number.

An interval number can be viewed as a special fuzzy number whose membership function takes value 1 over the interval, and 0 anywhere else. It is clear that the above three operations of intervals are equivalent to the operations of addition, subtraction and scalar multiplication of fuzzy numbers via the extension principle. Rommelfanger, Hanscheck and Wolf investigated the interval programming problem as a fuzzy programming problem.

Ishibuchi and Tanaka suggested an order relation $\preceq$ between two intervals as follows.

Definition 8.2: If intervals a and b are two profit intervals, the order relation $\preceq$ between a and b is defined as

$$a \preceq b \text{ if and only if } \underline{a} \leq \underline{b} \text{ and } \overline{a} \leq \overline{b}; \tag{8.9}$$

$$a \prec b \text{ if and only if } a \preceq b \text{ and } a \neq b. \tag{8.10}$$

For describing the interval inequality relation in detail, we give three new concepts, as follows:

Definition 8.3: For any two interval numbers $a = [\underline{a}, \overline{a}]$ and $b = [\underline{b}, \overline{b}]$, there is an interval inequality relation $a \leq b$ between the two interval numbers a and b if and only if $m(a) \leq m(b)$. Furthermore, if $\overline{a} \leq \underline{b}$, we say the interval inequality relation $a \leq b$ between a and b is optimistic satisfactory; if $\overline{a} > \underline{b}$, we say the interval inequality relation $a \leq b$ between a and b is pessimistic satisfactory.

Definition 8.4: For any two interval numbers $a = [\underline{a}, \overline{a}]$ and $b = [\underline{b}, \overline{b}]$, if the interval inequality relation between them is pessimistic satisfactory, the pessimistic satisfaction index of the interval inequality relation $a \leq b$ can be defined as

$$PSD(a \leq b) = 1 + \frac{\underline{b} - \overline{a}}{w(a) + w(b)} \tag{8.11}$$

Definition 8.5: For any two interval numbers $a = [\underline{a}, \overline{a}]$ and $b = [\underline{b}, \overline{b}]$, if the interval inequality relation between them is optimistic satisfactory, the optimistic satisfaction index of the interval inequality relation $a \leq b$ can be defined as

$$OSD(a \leq b) = \frac{\underline{b} - \overline{a}}{w(a) + w(b)} \tag{8.12}$$

Remark 8.1: According to definitions of the pessimistic satisfactory and the optimistic satisfaction indices, we can see that the domain of the pessimistic satisfaction index can be $[0, 1)$ and the domain of the optimistic satisfaction index can be $[0, \infty)$. The larger the values of the pessimistic satisfaction and optimistic satisfaction indices are, the larger are the satisfaction degrees of the interval relations.

8.3 The Expected Return Intervals of Securities

It is well known that future returns of securities cannot be accurately predicted in any emerging securities market. Traditionally, researchers consider the arithmetic mean of historical returns as the expected return of the security. So the expected return of the security is a crisp value in this way. However for this technique, two main problems need to be solved:

(1) If the time horizon of the historical data of a security is very long, the influence of the earlier historical data is the same as that of the recent data. However, recent data of a security most often indicate that the performance of a corporation is more important in recent data than in the earlier historical data.

(2) If the historical data of a security are not enough, one cannot accurately estimate the statistical parameters, due to data scarcity.

Considering these two problems, perhaps it is a good idea to consider the expected return of a security as an interval number, rather than a crisp value, based on the arithmetic mean of historical data. Investors may make use of a corporation's financial reports and the security's historical data to determine the expected return interval's range.

To determine the range of change in expected returns of securities, we will consider the following three factors:

(1) Arithmetic mean: Although arithmetic means of returns of securities should not be expressed as expected returns directly, they are a good approximation. Denote the arithmetic mean return factor as r_a, which can be calculated with historical data.

(2) Historical return tendency: If recent returns of a security have been increasing, we can believe that the expected return of the security is greater than the arithmetic mean based on historical data. However, if recent returns of a security have been declining, we can assume that the expected return of the security is smaller than the arithmetic mean based on historical data. Denote the historical return tendency factor as r_h, r_h, which reflects the tendency of the return on the security. We can use the arithmetic mean of recent returns as r_h.

(3) Forecast of future returns of a security: The third factor influencing the expected return of a security is its estimated future returns. Based on the financial reports of a corporation, if we believe that the returns on this corporation's stock will increase, then the expected returns of this security should be larger than the arithmetic mean based on historical data. On the contrary, if we think that returns of this corporation's stock will decrease in future, the expected return of this security will be smaller than the arithmetic mean. Denote the forecast return factor as r_f. Computation of derivation of r_f requires some forecasts based on the financial reports and experts' individual experiences.

Based on the above three factors, we can derive lower and upper limits of the expected return of the security. We can put the minimum of the three factors, r_a, r_h and r_f, as the lower limit of the expected return, while we can put the maximum values of the three factors r_a, r_h and r_f as the upper limit of the expected return of the security.

8.4 The Interval Programming Models for Portfolio Selection

Assume that an investor wants to allocate his wealth among n risky assets offering random rates of returns and a risk-free asset offering a fixed rate of return. We introduce some notations as follows.

$\tilde{r}_j$: the expected rate of return interval of risky asset j $(j = 1, 2, \cdots, n)$;

r_{n+1}: the rate of return of risk-free asset $n + 1$;

x_i: the proportion of the total investment devoted to risky asset i $(i = 1, 2, \cdots, n)$ or risk-free asset $n + 1$;

x_i^0: the proportion of the risky asset i $(i = 1, 2, \cdots, n)$ or risk-free asset $n + 1$ owned by the investor;

r_{tj}: the historical rate of return of risky asset j $(j = 1, 2, \cdots, n)$, t $(t = 1, 2, \cdots, T)$;

k_i: the rate of transaction costs for the asset i $(i = 1, 2, \cdots, n + 1)$;

u_i: the upper bound of proportion of the total investment devoted to risky asset i $(i = 1, 2, \cdots, n)$ or risk-free asset $n + 1$.

We use a V shape function to express the transaction costs, so the transaction costs of the asset i $(i = 1, 2, \cdots, n + 1)$ can be denoted by

$$C_i(x_i) = k_i |x_i - x_i^0|. \tag{8.13}$$

So the total transaction costs of the portfolio $x = (x_1, x_2, \cdots, x_n, x_{n+1})$ can be denoted by

$$C(x) = \sum_{i=1}^{n+1} C_i(x_i) = \sum_{i=1}^{n+1} k_i |x_i - x_i^0|. \tag{8.14}$$

Denote

$$r_{aj} = \frac{1}{T} \sum_{t=1}^{T} r_{tj}. \tag{8.15}$$

The uncertain expected return of the risky asset j $(j = 1, 2, \cdots, n)$ can be represented as the following interval number:

$$\tilde{r}_j = [\underline{r}_j, \overline{r}_j] = [\min\{r_{aj}, r_{hj}, r_{fj}\}, \max\{r_{aj}, r_{hj}, r_{fj}\}], \tag{8.16}$$

where r_{aj} is the arithmetic mean factor of risky asset j, r_{hj} is the historical return tendency factor of risky asset j and r_{fj} is the forecast return factor of risky asset j. They can be derived by using the above method.

So the expected return interval of portfolio $x = (x_1, x_2, \cdots, x_{n+1})$ in the future can be represented as

$$\hat{r}(x) = \sum_{j=1}^{n} \tilde{r}_j x_j + r_{n+1} x_{n+1}. \tag{8.17}$$

After removing the transaction costs, the net expected return interval of portfolio $x = (x_1, x_2, \cdots, x_{n+1})$ can be represented as

$$\tilde{r}(x) = \sum_{j=1}^{n} \tilde{r}_j x_j + r_{n+1} x_{n+1} - \sum_{i=1}^{n+1} k_i |x_i - x_i^0|. \tag{8.18}$$

If the expected returns of securities are crisp values, the semi-absolute deviation of the return of portfolio $x = (x_1, x_2, \cdots, x_{n+1})$ below the expected return at the past period t, $t = 1, 2, \cdots, T$ can be represented as

$$w_t(x) = |\min\{0, \sum_{j=1}^{n}(r_{tj} - r_j)x_j\}| = \max\{0, \sum_{j=1}^{n}(r_j - r_{tj})x_j\} \qquad (8.19)$$

where r_j is the expected return of security j.

Because the expected returns on securities are considered as interval numbers, we may consider the semi-absolute deviation of the rates of return of portfolio $x = (x_1, x_2, \cdots, x_{n+1})$ below the expected return over all the past periods as an interval number too.

Since the expected return interval of portfolio $x = (x_1, x_2, \cdots, x_{n+1})$ is

$$\hat{r}(x) = [\sum_{j=1}^{n}\underline{r}_j x_j + r_{n+1}x_{n+1}, \sum_{j=1}^{n}\overline{r}_j x_j + r_{n+1}x_{n+1}], \qquad (8.20)$$

we can get the semi-absolute deviation interval of return of portfolio $x = (x_1, x_2, \cdots, x_{n+1})$, below the expected return over the past period t, $t = 1, 2, \cdots, T$. It can be represented as

$$\tilde{w}_t(x) = [\max\{0, \sum_{j=1}^{n}(\underline{r}_j - r_{tj})x_j\}, \max\{0, \sum_{j=1}^{n}(\overline{r}_j - r_{tj})x_j\}]. \qquad (8.21)$$

Then the average value of the semi-absolute deviation interval of return of portfolio $x = (x_1, x_2, \cdots, x_{n+1})$, below the uncertain expected return over all the past periods, can be represented as

$$\begin{aligned}
\tilde{w}(x) &= \tfrac{1}{T}\sum_{t=1}^{T}\tilde{w}_t(x) \\
&= \tfrac{1}{T}\sum_{t=1}^{T}[\max\{0, \sum_{j=1}^{n}(\underline{r}_j - r_{tj})x_j\}, \max\{0, \sum_{j=1}^{n}(\overline{r}_j - r_{tj})x_j\}]
\end{aligned} \qquad (8.22)$$

We use $\tilde{w}(x)$ to measure the risk of portfolio x. Suppose that the investor wants to maximize the return of a portfolio after removing the transaction costs within some given level of risk. If the risk tolerance interval $\tilde{w}$ is given, the mathematical formulation of the portfolio selection problem is

$$\begin{aligned}
(ILP1) \quad \max \quad &\tilde{r}(x) = \sum_{j=1}^{n}\tilde{r}_j x_j + r_{n+1}x_{n+1} - \sum_{i=1}^{n+1} k_i|x_i - x_i^0| \\
\text{s.t.} \quad &\tilde{w}(x) \leq [\underline{w}, \overline{w}], \\
&\sum_{j=1}^{n+1} x_j = 1, \\
&0 \leq x_j \leq u_j, j = 1, 2, \cdots, n+1
\end{aligned}$$

where $\underline{w}$ and $\overline{w}$ are two given constants, $\underline{w}$ represents the pessimistic tolerated risk level, and $\overline{w}$ represents the optimistic tolerated risk level.

(ILP1) is an optimization problem with interval coefficients and, therefore, techniques of classical linear programming can not be applied unless the above interval optimization problem is reduced to a standard linear programming structure. In the following, we perform this conversion.

We introduce the order relation $\preceq$ in the interval objective function of (ILP1). Based on the concepts of pessimistic and optimistic satisfaction indices proposed by us in Section 2, the interval inequality relation $\tilde{w}(x) \leq [\underline{w}, \overline{w}]$ in (ILP1) is sure to be expressed by one of the two crisp equalities. The two crisp equivalent equalities of the interval constraint condition $\tilde{w}(x) \leq [\underline{w}, \overline{w}]$ can be represented as follows:

$$PSD(\tilde{w}(x) \leq [\underline{w}, \overline{w}]) = \alpha \qquad (8.23)$$

and

$$OSD(\tilde{w}(x) \leq [\underline{w}, \overline{w}]) = \beta. \qquad (8.24)$$

Then the interval linear programming problem (ILP1) can be decomposed into two interval linear programming problems in which the objective functions are interval numbers and the constraint conditions are crisp equalities and inequalities. The two interval objective function linear programming problems are represented as follows:

$$(PO1) \ \max_{\preceq} \ \tilde{r}(x) = \sum_{j=1}^{n} \tilde{r}_j x_j + r_{n+1} x_{n+1} - \sum_{i=1}^{n+1} k_i |x_i - x_i^0|$$
$$\text{s.t.} \ PSD(\tilde{w}(x) \leq [\underline{w}, \overline{w}]) = \alpha,$$
$$\sum_{j=1}^{n+1} x_j = 1,$$
$$x_j \geq 0, j = 1, 2, \cdots, n+1,$$
$$\alpha \in [0, 1).$$

where α is given by the investor.

$$(PS1) \ \max_{\preceq} \ \tilde{r}(x) = \sum_{j=1}^{n} \tilde{r}_j x_j + r_{n+1} x_{n+1} - \sum_{i=1}^{n+1} k_i |x_i - x_i^0|$$
$$\text{s.t.} \ OSD(\tilde{w}(x) \leq [\underline{w}, \overline{w}]) = \beta,$$
$$\sum_{j=1}^{n+1} x_j = 1,$$
$$x_j \geq 0, j = 1, 2, \cdots, n+1,$$
$$\beta \in [0, \infty).$$

where β is given by the investor.

We can see that the constraint conditions of (PO1) are stricter than those of (PS1). Hence, we can get an optimistic investment strategy by solving (PO1), and a pessimistic investment strategy by solving (PS1).

Denote F_1 as the feasible set of (PO1) and F_2 as the feasible set of (PS1).

Definition 8.6: $x \in F_1$ is a satisfactory solution of (PO1) if and only if there is no other $x' \in F_1$ such that $\tilde{r}(x) \prec \tilde{r}(x')$; $x \in F_2$ is a satisfactory solution of (PS1) if and only if there is no other $x' \in F_2$ such that $\tilde{r}(x) \prec \tilde{r}(x')$.

By Definition 8.6, the satisfactory solution of (PO1) is equivalent to the non-inferior solution set of the following bi-objective programming problem:

$$(BLP1) \quad \max \quad \sum_{j=1}^{n} \underline{r}_j x_j + r_{n+1} x_{n+1} - \sum_{i=1}^{n+1} k_i |x_i - x_i^0|$$

$$\max \quad \sum_{j=1}^{n} \overline{r}_j x_j + r_{n+1} x_{n+1} - \sum_{i=1}^{n+1} k_i |x_i - x_i^0|$$

$$\text{s.t. } PSD(\tilde{w}(x) \le [\underline{w}, \overline{w}]) = \alpha,$$

$$\sum_{j=1}^{n+1} x_j = 1,$$

$$x_j \ge 0, j = 1, 2, \cdots, n+1,$$

$$\alpha \in [0, 1).$$

The satisfactory solution of (PS1) is equivalent to the non-inferior solution set of the following bi-objective programming problem:

$$(BLP2) \quad \max \quad \sum_{j=1}^{n} \underline{r}_j x_j + r_{n+1} x_{n+1} - \sum_{i=1}^{n+1} k_i |x_i - x_i^0|$$

$$\max \quad \sum_{j=1}^{n} \overline{r}_j x_j + r_{n+1} x_{n+1} - \sum_{i=1}^{n+1} k_i |x_i - x_i^0|$$

$$\text{s.t. } OSD(\tilde{w}(x) \le [\underline{w}, \overline{w}]) = \beta,$$

$$\sum_{j=1}^{n+1} x_j = 1,$$

$$x_j \ge 0, j = 1, 2, \cdots, n+1,$$

$$\beta \in [0, \infty).$$

By the multi-objective programming theory, the non-inferior solution to (BLP1) can be generated by solving the following parametric linear programming problem:

$$(PLP1) \quad \max \quad \sum_{j=1}^{n} [\lambda \underline{r}_j + (1 - \lambda) \overline{r}_j] x_j + r_{n+1} x_{n+1} - \sum_{i=1}^{n+1} k_i |x_i - x_i^0|$$

$$\text{s.t. } PSD(\tilde{w}(x) \le [\underline{w}, \overline{w}]) = \alpha,$$

$$\sum_{j=1}^{n+1} x_j = 1,$$

$$x_j \ge 0, j = 1, 2, \cdots, n+1,$$

$$\alpha \in [0, 1).$$

The non-inferior solution to (BLP2) can be generated by solving the following parametric linear programming problem:

$$(PLP2) \max \ \sum_{j=1}^{n} [\lambda \underline{r}_j + (1 - \lambda)\overline{r}_j]x_j + r_{n+1}x_{n+1} - \sum_{i=1}^{n+1} k_i|x_i - x_i^0|$$

$$\text{s.t. } OSD(\tilde{w}(x) \le [\underline{w}, \overline{w}]) = \beta,$$

$$\sum_{j=1}^{n+1} x_j = 1,$$

$$x_j \ge 0, j = 1, 2, \cdots, n + 1,$$

$$\beta \in [0, \infty).$$

Introducing the concrete form of $PSD(\tilde{w}(x) \le [\underline{w}, \overline{w}])$, (PLP1) may be rewritten as follows:

$$(PLP3) \max \ \sum_{j=1}^{n} [\lambda \underline{r}_j + (1 - \lambda)\overline{r}_j]x_j + r_{n+1}x_{n+1} - \sum_{i=1}^{n+1} k_i|x_i - x_i^0|$$

$$\text{s.t. } \frac{1}{T} \sum_{t=1}^{T} [(1 + \alpha) \max\{0, \sum_{j=1}^{n} (\overline{r}_j - r_{tj})x_j\}$$

$$+ (1 - \alpha) \max\{0, \sum_{j=1}^{n} (\underline{r}_j - r_{tj})x_j\}]$$

$$= (1 - \alpha)\overline{w} + (1 + \alpha)\underline{w},$$

$$\sum_{j=1}^{n+1} x_j = 1,$$

$$x_j \ge 0, j = 1, 2, \cdots, n + 1,$$

$$\alpha \in [0, 1).$$

Introducing the concrete form of $OSD(\tilde{w}(x) \le [\underline{w}, \overline{w}])$, (PLP2) may be rewritten as follows:

$$(PLP4) \max \ \sum_{j=1}^{n} [\lambda \underline{r}_j + (1 - \lambda)\overline{r}_j]x_j + r_{n+1}x_{n+1} - \sum_{i=1}^{n+1} k_i|x_i - x_i^0|$$

$$\text{s.t. } \frac{1}{T} \sum_{t=1}^{T} [(2 + \beta) \max\{0, \sum_{j=1}^{n} (\overline{r}_j - r_{tj})x_j\}$$

$$- \beta \max\{0, \sum_{j=1}^{n} (\underline{r}_j - r_{tj})x_j\}] = (2 + \beta)\underline{w} - \beta\overline{w},$$

$$\sum_{j=1}^{n+1} x_j = 1,$$

$$x_j \ge 0, j = 1, 2, \cdots, n + 1,$$

$$\beta \in [0, \infty).$$

To solve (PLP3) and (PLP4), we consider the following transformation. Introducing a new variable x_{n+2} such that

$$x_{n+2} \ge \sum_{i=1}^{n+1} k_i|x_i - x_i^0|. \tag{8.25}$$

Let

$$d_i^+ = \frac{|x_i - x_i^0| + (x_i - x_i^0)}{2}; \tag{8.26}$$

$$d_i^- = \frac{|x_i - x_i^0| - (x_i - x_i^0)}{2}; \tag{8.27}$$

$$\underline{y}_t^+ = \frac{|\sum_{j=1}^{n}(r_{tj} - \underline{r}_j)x_j| + \sum_{j=1}^{n}(r_{tj} - \underline{r}_j)x_j}{2}; \tag{8.28}$$

$$\underline{y}_t^- = \frac{|\sum_{j=1}^{n}(r_{tj} - \underline{r}_j)x_j| - \sum_{j=1}^{n}(r_{tj} - \underline{r}_j)x_j}{2}; \tag{8.29}$$

$$\overline{y}_t^+ = \frac{|\sum_{j=1}^{n}(r_{tj} - \overline{r}_j)x_j| + \sum_{j=1}^{n}(r_{tj} - \overline{r}_j)x_j}{2}; \tag{8.30}$$

$$\overline{y}_t^- = \frac{|\sum_{j=1}^{n}(r_{tj} - \overline{r}_j)x_j| - \sum_{j=1}^{n}(r_{tj} - \overline{r}_j)x_j}{2}. \tag{8.31}$$

Thus, (PLP3) can be rewritten as follows:

$(PLP5)$ max $\sum_{j=1}^{n}[\lambda \underline{r}_j + (1 - \lambda)\overline{r}_j]x_j + r_{n+1}x_{n+1} - x_{n+2}$

s.t. $\frac{1}{T}\sum_{t=1}^{T}(1 + \alpha)\overline{y}_t^- + (1 - \alpha)\underline{y}_t^- = (1 - \alpha)\overline{w} + (1 + \alpha)\underline{w},$

$\sum_{j=1}^{n+1} k_j(d_i^+ + d_i^-) \leq x_{n+2},$

$\overline{y}_t^- + \sum_{j=1}^{n}(r_{tj} - \overline{r}_j)x_j \geq 0,$

$\underline{y}_t^- + \sum_{j=1}^{n}(r_{tj} - \underline{r}_j)x_j \geq 0,$

$d_i^+ - d_i^- = x_i - x_i^0, i = 1, 2, \cdots, n + 1,$

$\sum_{j=1}^{n+1} x_j = 1,$

$d_i^+ \geq 0, d_i^- \geq 0, i = 1, 2, \cdots, n + 1,$

$\underline{y}_t^- \geq 0, \overline{y}_t^- \geq 0, t = 1, 2, \cdots, T,$

$x_j \geq 0, j = 1, 2, \cdots, n + 1,$

$\alpha \in [0, 1).$

(PLP4) can be rewritten as follows:

$$(PLP6) \max \sum_{j=1}^{n} [\lambda \underline{r}_j + (1-\lambda)\overline{r}_j]x_j + r_{n+1}x_{n+1} - x_{n+2}$$

$$\text{s.t. } \frac{1}{T}\sum_{t=1}^{T}(2+\beta)\overline{y}_t^- - \beta \underline{y}_t^- = (2+\beta)\underline{w} - \beta\overline{w},$$

$$\sum_{j=1}^{n+1} k_j(d_i^+ + d_i^-) \le x_{n+2},$$

$$\overline{y}_t^- + \sum_{j=1}^{n}(r_{tj} - \overline{r}_j)x_j \ge 0,$$

$$\underline{y}_t^- + \sum_{j=1}^{n}(r_{tj} - \underline{r}_j)x_j \ge 0,$$

$$d_i^+ - d_i^- = x_i - x_i^0, i = 1, 2, \cdots, n+1,$$

$$\sum_{j=1}^{n+1} x_j = 1,$$

$$d_i^+ \ge 0, d_i^- \ge 0, i = 1, 2, \cdots, n+1,$$

$$\underline{y}_t^- \ge 0, \overline{y}_t^- \ge 0, t = 1, 2, \cdots, T,$$

$$x_j \ge 0, j = 1, 2, \cdots, n+1,$$

$$\beta \in [0, \infty).$$

(PLP5) and (PLP6) are two standard linear programming problems. One can use several algorithms of linear programming to solve them efficiently, for example, the simplex method. So we can solve the original portfolio selection problem (ILP1) by solving (PLP5) and (PLP6).

8.5 Numerical Example

In this section, we suppose that an investor chooses twelve componental stocks in the Shanghai 30 index and a risk-less asset for his investment. We use a kind of saving account as the risk-less asset and the term of the saving account is three months. So the rate of return of the risk-less asset is 0.0014 per month. We collect historical data of the twelve stocks from January, 1999 to December, 2002. The data are downloaded from the web site www.stockstar.com. Then we use one month as a period to obtain historical rates of return, during forty-eight periods. The names of the twelve stocks are given in Table 1.

Because the securities markets in mainland China are very young, the arithmetical methods do not produce good estimates of the actual returns that the investor will receive in the future. According to our method in Section 3, we can obtain the expected rate of return interval of each stock. The intervals are given in Table 2.

Suppose the investor stipulates risk level interval $\tilde{w} = [0.015, 0.040]$; by the method proposed in the above section, we can solve the portfolio selection problem by solving (PLP5) and (PLP6).

For the given risk level interval $\tilde{w}$, more optimistic portfolios can be generated by varying the values of the parameters λ and α in (PLP5); more

Table 8.1. The exchange codes and the names of the twelve stocks

600001	Handan Steel	600002	Qilu Petrochemical
600009	Shanghai Airdrome	600058	Longteng Technology
600068	Gezhou Dam	600072	Jiangnan Heavy Industry
600098	Guangzhou Holding	600100	Tsinghua Tongfang
600104	Shanghai Auto	600115	East Airways
600120	Zhejiang East	600631	First Department Store

Table 8.2. The expected rates of returns intervals

Exchange Code	600001	600002	600009
Return Interval	[0.0060, 0.0068]	[0.0062, 0.0069]	[0.0104, 0.0114]
Exchange Code	600058	600068	600072
Return Interval	[0.0231, 0.0238]	[0.0067, 0.0078]	[0.0089, 0.0098]
Exchange Code	600098	600100	600104
Return Interval	[0.0164, 0.0173]	[0.0261, 0.0268]	[0.0078, 0.0087]
Exchange Code	600115	600120	600631
Return Interval	[0.0156, 0.0167]	[0.0223, 0.0229]	[0.0120, 0.0128]

pessimistic portfolios can be generated by varying the values of the parameters λ and β in (PLP6).

The return intervals, the risk intervals and the values of parameters of optimistic portfolios are listed in Table 3. The optimistic portfolios are listed in Table 4. The return intervals, the risk intervals and the values of parameters of pessimistic portfolios are listed in Table 5. The pessimistic portfolios are listed in Table 6.

Table 8.3. The return intervals, the risk intervals and the values of parameters of optimistic portfolios

	Return Interval	Risk Interval	λ	α
Portfolio 1	[0.0145, 0.0149]	[0.0273, 0.0276]	0.60	0
Portfolio 2	[0.0140, 0.0145]	[0.0248, 0.0251]	0.50	0.2
Portfolio 3	[0.0106, 0.0110]	[0.0178, 0.0180]	0.30	0.8

The investor may choose his own investment strategy from the portfolios according to his attitude towards the securities' expected returns and the degree of portfolio risk with which he is comfortable. If the investor is not

Table 8.4. The allocation of Portfolio $1, 2, 3$

Exchange Code	600001	600002	600009	600058	600068
Portfolio 1	0.0000	0.0000	0.0000	0.0000	0.0000
Portfolio 2	0.0000	0.0000	0.0000	0.0730	0.0000
Portfolio 3	0.0000	0.0000	0.0000	0.0572	0.0000
Exchange Code	600072	600098	600100	600104	600115
Portfolio 1	0.0000	0.0000	0.4146	0.0000	0.0000
Portfolio 2	0.0000	0.0000	0.2886	0.0000	0.0000
Portfolio 3	0.0000	0.0000	0.1825	0.0000	0.0000
Exchange Code	600120	600631	Saving		
Portfolio 1	0.3078	0.0000	0.2776		
Portfolio 2	0.3610	0.0000	0.2774		
Portfolio 3	0.2938	0.0000	0.4665		

Table 8.5. The return intervals, the risk intervals and the values of parameters of pessimistic portfolios

	Return Interval	Risk Interval	λ	β
Portfolio 4	[0.0091, 0.0094]	[0.0148, 0.0150]	0.60	0
Portfolio 5	[0.0066, 0.0068]	[0.0144, 0.0147]	0.50	0.8
Portfolio 6	[0.0049, 0.0052]	[0.0132, 0.0138]	0.30	1.5

satisfied with any of these portfolios, he may obtain more by solving the two parametric linear programming problems (PLP5) and (PLP6).

8.6 Conclusion

An approach is presented for estimating intervals of rates of returns of securities. The semi-absolute deviation risk function is extended to an interval case. An interval semi-absolute deviation model with no short selling and no stock borrowing in a frictional market is proposed for portfolio selection. By introducing the concepts of pessimistic and optimistic satisfaction indices of the interval inequality relation, an approach to compare interval numbers is given. By using the approach, the interval semi-absolute deviation model can be converted into two parametric linear programming problems. One can find a satisfactory solution to the original problem by solving the corresponding parametric linear programming problems. An investor may choose a satisfactory investment strategy according to an optimistic or pessimistic attitude. The model is capable of helping the investor to find an efficient portfolio that is closest to his/her targets.

Table 8.6. The allocation of Portfolio $4, 5, 6$

Exchange Code	600001	600002	600009	600058	600068
Portfolio 4	0.0000	0.0000	0.0189	0.0472	0.0000
Portfolio 5	0.0000	0.0000	0.0000	0.0000	0.0000
Portfolio 6	0.0000	0.0000	0.0000	0.0000	0.0000
Exchange Code	600072	600098	600100	600104	600115
Portfolio 4	0.0090	0.0000	0.0839	0.0000	0.0000
Portfolio 5	0.0000	0.0000	0.2662	0.0000	0.0000
Portfolio 6	0.0000	0.0000	0.1850	0.0000	0.0000
Exchange Code	600120	600631	Saving		
Portfolio 4	0.3274	0.0000	0.5136		
Portfolio 5	0.0000	0.0000	0.7338		
Portfolio 6	0.0000	0.0000	0.8150		

9

Quadratic Programming Model with Interval Coefficients

9.1 Introduction

Markowitz's mean-variance model for portfolio selection, proposed in 1952, assumed that the stock market is frictionless. However, the real stock market is imperfect. Though the probability theory is a major tool used for analyzing uncertainty in financial markets, it can not describe the uncertainty completely since there are many other uncertain factors that differ from the random ones. As an alternative tool, the fuzzy set theory is gradually being used in this area. In this chapter, we present a model where expected return and risk (variance) are treated as interval numbers.

9.2 Crisp Model and Algorithm

We consider a financial market with n risky assets offering random rates of return and a risk-free asset offering a fixed return rate. An investor allocates his/her wealth among the risky assets and the risk-free asset. A portfolio can be described by the following scalar-values:

x_i, the proportion invested in the risky asset i, $i = 1, \cdots, n$;

x_{n+1}, the proportion invested in the risk-less asset;

r_i, the random rate of return of risky asset i;

R_{n+1}, the return rate of the risk-free asset ;

$R_i = E(r_i)$, the expected rate of return of the risky asset i;

$\sigma_{ij} = cov(r_i, r_j)$, the covariance between r_i and r_j, $i, j = 1, \cdots, n$, and the variance-covariance matrix $(\sigma_{ij})_{n*n}$ which is semi-positive definite;

c_i, the transaction cost of the risky asset i, $i = 1, \cdots, n$;

k_i, the constant cost per change in a proportion of the risky asset i, $k_i \geq 0, i = 1, \cdots, n$.

In this chapter, the transaction cost c_i is assumed to be a V-shaped function of the difference between a given portfolio $x^0 = (x_1^0, x_2^0, \cdots, x_n^0, x_{n+1}^0)^T$

and a new portfolio $x = (x_1, \cdots, x_n, x_{n+1})$, and is incorporated explicitly into the portfolio return. Thus, the transaction cost of risky asset i can be expressed as $c_i = k_i|x_i - x_i^0|$, $i = 1, \cdots, n$, and the total transaction cost is

$$\sum_{i=1}^{n} c_i = \sum_{i=1}^{n} k_i|x_i - x_i^0| \tag{9.1}$$

the expected return and variance of portfolio $x = (x_1, x_2, \cdots, x_n, x_{n+1})$ are

$$R(x) = \sum_{i=1}^{n+1} R_i x_i - \sum_{i=1}^{n} k_i|x_i - x_i^0| \tag{9.2}$$

and

$$\sigma^2(x) = \sum_{i=1}^{n} \sum_{j=1}^{n} \sigma_{ij} x_i x_j \tag{9.3}$$

respectively. For a new investor, it can be taken that $x_i^0 = 0$, $i = 1, \cdots, n, n+1$.

The investor expects to maximize the expected portfolio return $R(x)$ and to minimize the risk $\sigma^2(x)$. Mathematically, the portfolio selection problem can be formulated as the following bi-objective problem:

$$max \ R(x) = \sum_{i=1}^{n+1} R_i x_i - \sum_{i=1}^{n} k_i|x_i - x_i^0|$$

$$min \ \sigma^2(x) = \sum_{i=1}^{n} \sum_{j=1}^{n} \sigma_{ij} x_i x_j$$

$$s.t. \ \sum_{i=1}^{n+1} x_i = 1$$

$$x_i \geq 0, \ i = 1, \cdots, n+1. \tag{9.4}$$

According to the multi-objective programming theory, the efficient solution to above problem can be obtained by solving the following problem:

$$min \ \sigma^2(x) - \alpha R(x)$$

$$s.t. \ \sum_{i=1}^{n} x_i = 1$$

$$x_i \geq 0, \ i = 1, \cdots, n+1, \tag{9.5}$$

where $\alpha \in [0, +\infty)$.

Theorem 1 $x^* = (x_1^*, \cdots, x_{n+1}^*)^T$ is an optimal solution to (5) if and only if there exists $(y_1^*, \cdots, y_n^*)^T$ such that $(x_1^*, \cdots, x_{n+1}^*, y_1^*, \cdots, y_n^*)^T$ is an optimal solution to the following problem (6):

$$min \sum_{i=1}^{n} \sum_{j=1}^{n} \sigma_{ij} x_i x_j - \alpha [\sum_{i=1}^{n+1} R_i x_i - \sum_{i=1}^{n} k_i y_i]$$

$$s.t. \sum_{i=1}^{n+1} x_i = 1$$

$$y_i + x_i - x_i^0 \geq 0, i = 1, \cdots, n$$

$$y_i - x_i + x_i^0 \geq 0, i = 1, \cdots, n$$

$$x_j \geq 0, j = 1, \cdots, n+1, \tag{9.6}$$

where $\alpha \in [0, +\infty)$.

The convex quadratic programming problem can be solved by several efficient numerical methods.

9.3 The Model with Interval Coefficients and Its Extension

In a fuzzy environment, the expected return and risk can not be predicted accurately, so the investor usually makes his portfolio decision according to his experience and his economic sense. Based on this fact, we propose a fuzzy model under the assumption that the expected return and risk are interval numbers. One can refer to Alefeld and Herzberger for a detailed discussion on interval number operations.

Denote the fuzzy expected return and covariance as the following interval numbers:

$$\widetilde{R_i} = [R_i - \delta_{il}, R_i + \delta_{ir}],$$
$$\widetilde{\sigma_{ij}} = [\sigma_{ij} - \delta_{ijl}, \sigma_{ij} + \delta_{ijr}], \tag{9.7}$$

where $\delta_{il}, \delta_{ir}, \sigma_{ijl}, \delta_{ijr}$ are positive constants given by the investor such that $\delta_{il} \leq R_i, \delta_{ijl} \leq \sigma_{ij}$. Then, the fuzzy expected return and fuzzy risk are defined by

$$\widetilde{R}(x) = \sum_{i=1}^{n+1} \widetilde{R_i} x_i - \sum_{i=1}^{n} k_i |x_i - x_i^0| \tag{9.8}$$

and

$$\widetilde{\sigma^2}(x) = \sum_{i=1}^{n} \sum_{j=1}^{n} \widetilde{\sigma_{ij}} x_i x_j. \tag{9.9}$$

Since $x_i \geq 0$, we have fuzzy numbers

$$\widetilde{R}(x) = [R(x) - \delta_{RL}(x), R(x) + \delta_{RR}(x)],$$
$$\widetilde{\sigma^2}(x) = [\sigma^2(x) - \delta_{VL}(x), \sigma^2(x) + \delta_{VR}(x)], \tag{9.10}$$

where

$$R(x) - \delta_{RL}(x) = \sum_{i=1}^{n+1}(R_i - \delta_{il})x_i - \sum_{i=1}^{n} k_i|x_i - x_i^0|,$$

$$R(x) + \delta_{RR}(x) = \sum_{i=1}^{n+1}(R_i + \delta_{ir})x_i - \sum_{i=1}^{n} k_i|x_i - x_i^0|,$$

$$\sigma^2(x) - \delta_{VL}(x) = \sum_{i=1}^{n}\sum_{j=1}^{n}(\sigma_{ij} - \delta_{ijl})x_ix_j,$$

$$\sigma^2(x) + \delta_{VR}(x) = \sum_{i=1}^{n}\sum_{j=1}^{n}(\sigma_{ij} + \delta_{ijr})x_ix_j.$$

Corresponding to the optimization problem, the fuzzy optimization investment problem becomes:

$$min\ \widetilde{\sigma^2}(x) - \alpha\widetilde{R}(x)$$
$$s.t.\ \sum_{i=1}^{n+1} x_i = 1$$
$$x_i \geq 0, i = 1, \cdots, n. \tag{9.11}$$

Denote

$$F_l(x) = \sigma^2(x) - \delta_{VL}(x) - \alpha(R + \delta_{RR}(x)),$$
$$F_r(x) = \sigma^2(x) + \delta_{VR}(x) - \alpha(R - \delta_{RL}(x)),$$
$$F(x) = \sigma^2(x) - \alpha R(x),$$

then $\widetilde{\sigma^2}(x) - \alpha\widetilde{R}(x) = [F_l(x), F_r(x)]$.

From the above problem, we construct the following three models.

Model 1

$$min\ F_l(x)$$
$$s.t.\ \sum_{i=1}^{n+1} x_i = 1$$
$$x_i \geq 0, i = 1, \cdots, n. \tag{9.12}$$

From the objective function $F_l(x)$, we can observe that the investor estimates the return and risk optimistically, and aims to optimize the total objective in this case.

Model 2

$$min \ F_r(x)$$
$$s.t. \ \sum_{i=1}^{n+1} x_i = 1$$
$$x_i \geq 0, i = 1, \cdots, n \tag{9.13}$$

In contrast to model 1, the investor estimates the return and risk of risky assets pessimistically, and the investor aims to optimize the total objective based on his estimation.

Model 3

$$min \ F(x) = \lambda F_r(x) + (1 - \lambda) F_l(x)$$
$$s.t. \ \sum_{i=1}^{n+1} x_i = 1$$
$$x_i \geq 0, i = 1, \cdots, n, \tag{9.14}$$

where $\lambda \in (0, 1)$.

This model covers the scenario where the investor makes his portfolio selection neither too optimistically nor too pessimistically; to some extent he is optimistic, but to some extent he is pessimistic also.

Remark 9.1 This section provides three models for portfolio selection. The investor selects a different model on the basis of different estimations of return and risk.

Remark 9.2 The investor's subjective factor has an important impact on portfolio selection since fuzzy inputs $\widetilde{R}_i$ and $\widetilde{\sigma}_{ij}$ influence the output and the investor selects his portfolio model either pessimistically or optimistically.

9.4 Numerical Example

To illustrate the approach presented in the previous section, we consider a portfolio selection problem with three risky assets and a risk-free asset.

In the crisp case, the vector of their expected rates of return is:

$$(R_1, R_2, R_3, R_4) = (0.066, 0.0616, 0.055, 0.050)^T$$

and the covariance matrix of the return rates of three risky assets is:

$$(\sigma_{ij}) = \begin{pmatrix} 0.062 & 0.0151 & 0.0130 \\ 0.0151 & 0.0534 & 0.0126 \\ 0.0130 & 0.0126 & 0.0401 \end{pmatrix}$$

Let $k_i = 0.001, x_i^0 = 0$ for $i = 1, 2, 3$.

The efficient frontier of the crisp model is shown in Fig 9.1(in the figure, return and risk are the expected return and the variance of the portfolio respectively).

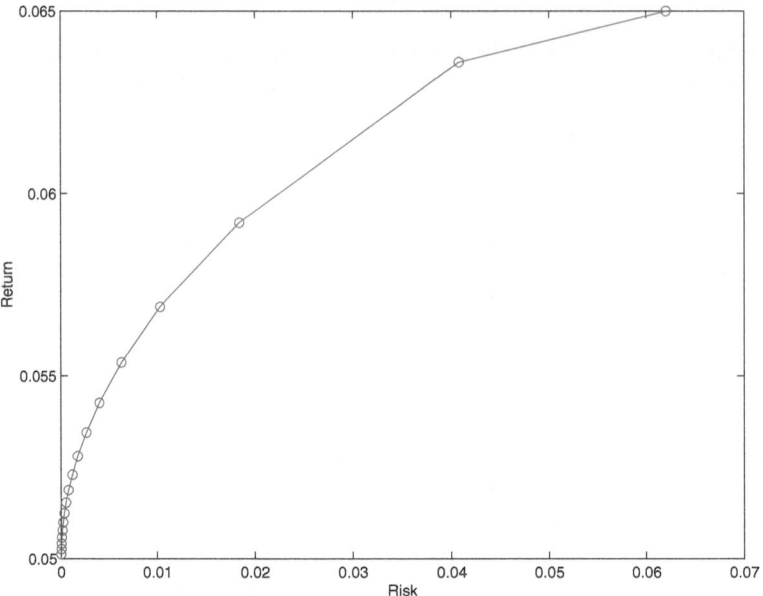

Fig. 9.1. Efficient frontier of the crisp model

Now, we consider the fuzzy case that the return and risk are given in the form of interval numbers by the investor as follows:

$$\widetilde{R_1} = [0.0540, 0.0726] \quad \widetilde{R_2} = [0.0544, 0.0678]$$
$$\widetilde{R_3} = [0.0495, 0.0605] \quad \widetilde{R_4} = [0.0500, 0.0500]$$
$$\widetilde{\sigma_{11}} = [0.0560, 0.0682] \quad \widetilde{\sigma_{12}} = [0.0136, 0.0166]$$
$$\widetilde{\sigma_{13}} = [0.0117, 0.0143] \quad \widetilde{\sigma_{22}} = [0.0481, 0.0587]$$
$$\widetilde{\sigma_{23}} = [0.0113, 0.0139] \quad \widetilde{\sigma_{33}} = [0.0361, 0.0441]$$

The efficient frontiers of model 1-3 in the fuzzy case are shown in Figs 9.2-9.4.

From the above discussion and Figs. 9.1-9.4, we can observe that the investor's subjective outlook has a great impact on his portfolio selection. Given a certain risk level, the expected return in the optimistic case is the highest of the four models; in the pessimistic case, it is the lowest; and the returns both in the crisp case and the situation between the optimistic case and pessimistic case are between the former two cases. For example, we can observe from Figures 1-4 that given a risk level 0.02, the return in the crisp case is

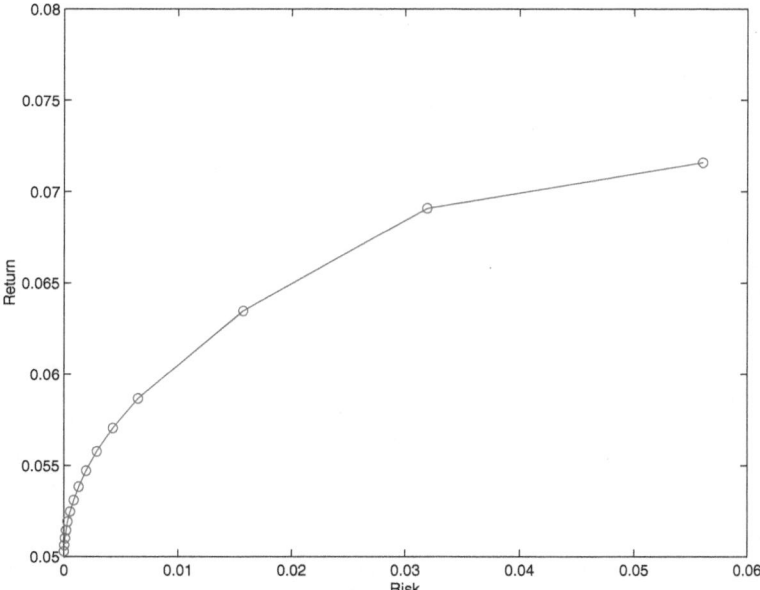

Fig. 9.2. Efficient frontier of Model 1

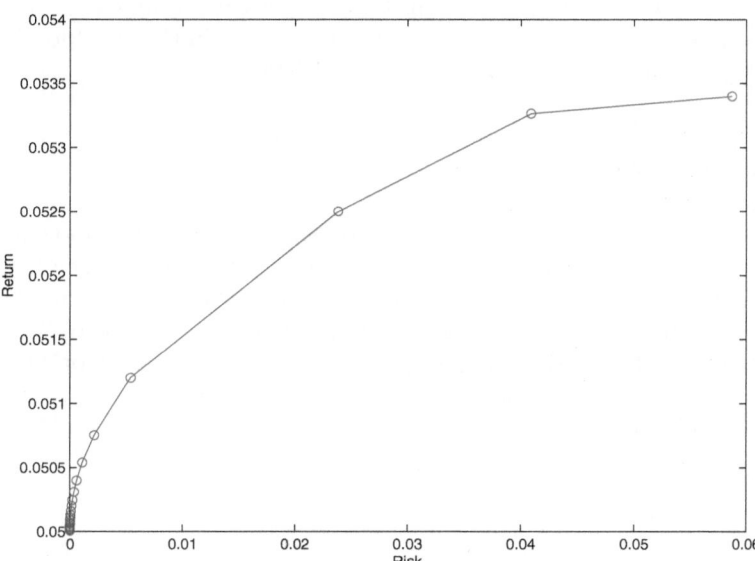

Fig. 9.3. Efficient frontier of Model 2

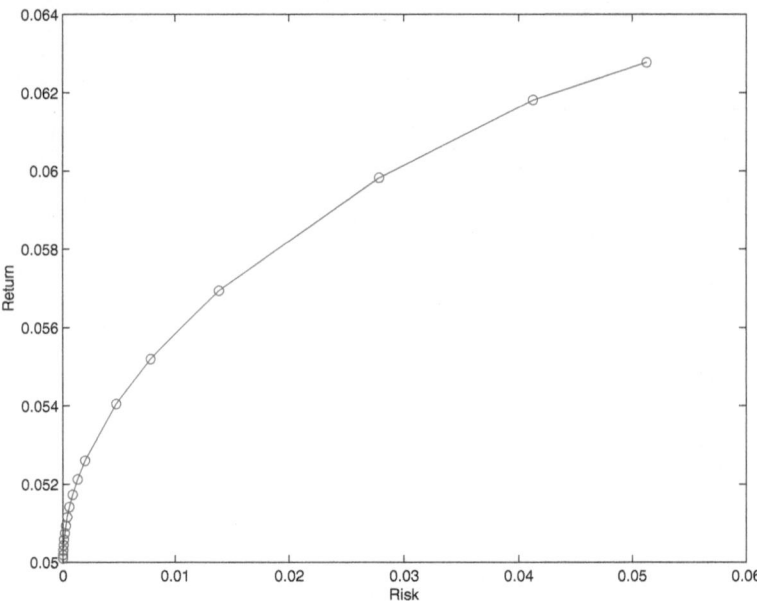

Fig. 9.4. Efficient frontier of Model 3

about 0.059 ; the return in the optimistic case is greater than 0.063, the return in the pessimistic case is lower than 0.053, the return in the situation between the optimistic and pessimistic cases is about 0.06.

9.5 Conclusion

In this section, we first propose a crisp non-smooth model for portfolio selection with transaction costs. We show an efficient approach that can transform this model into a quadratic programming problem. Next, we extend the crisp model to the fuzzy case where the return and risk are assumed to be interval numbers. Three interesting models, which can reflect the investor's objective opinion, are derived from this extended model. Finally, we give a numerical example to illustrate our methods.

Part IV

Portfolio Selection Models with Possibility
Distribution

Tanaka and Guo's Model with Exponential Possibility Distributions

10.1 Introduction

In the possibility theory (proposed by Zadeh and advanced by Dubois and Prade), fuzzy variables are associated with possibility distributions while the way random variables are associated with probability distributions. Possibility distributions are described as normal convex fuzzy sets, such as LR fuzzy numbers, quadratic and exponential functions. The theory of exponential possibility distributions has been proposed and applied to possibilistic data analysis. As an application of possibility theory to portfolio analysis, possibility portfolio selection models were initially proposed in Tanaka and Guo (1999) where portfolio models are based on exponential possibility distributions, rather than the mean-variance form in Markowitz's model. Although there are some similarities between Markowitz's model and possibility portfolio selection models, these two kinds of models analyze the security data in very different ways. Markowitz's model regards the portfolio selection as a probability phenomenon so that it minimizes the variance of portfolio return, subject to a given average return. On the contrary, possibility models based on possibility distributions reflect portfolio experts' knowledge, which is characterized by the given possibility grades to security data. The basic assumption for using Markowitz's model is that the situation of stock markets in future can be correctly reflected by security data of the past, that is, the mean and covariance of securities in future are similar to the past number. It is hard to justify this kind of assumption in the ever-changing stock markets. On the other hand, possibility portfolio models integrate the past security data and experts' judgement to catch variations of stock markets more plausibly. Because experts' knowledge is very valuable for predicting the future state of stock markets, it is reasonable that possibility portfolio models are useful in the real investment world.

The upper and lower possibility distributions have some similarities to upper and lower approximations in rough set theory. Multi-source knowledge from multiple experts is represented by a set of exponential possibility distributions. Based on the consistency index defined by the possibility measure of each pair of possibility distributions, Tanaka and Guo (2003) proposed a fusion model to integrate multiple possibility distributions into a new one representing a refined knowledge.

10.2 Possibility Distributions in Portfolio Selection Problems

Give data (r_i, h_i) $(i = 1, 2, \cdots, m)$ where $r_i = (r_{1i}, \cdots, r_{ni})^T$ is a vector of returns of n securities i at the ith period and h_i are an associated possibility grades assigned by expert knowledge to reflect a degree of similarity between the future state of stock markets and the state of the ith sample. These grades h_i are assumed to be expressed by a possibility distribution A, defined as

$$\pi_A(R) = \exp\{-(R-a)^T D_A^{-1}(R-a)\} = (a, D_A)_e, \tag{10.1}$$

where a is a center vector and D_A is a symmetric positive definite matrix, denoted as $D_A > 0$

Given the data, the problem is to determine an exponential possibility distribution, i.e., a center vector a and a symmetric positive definite matrix D_A. According to two different viewpoints, two kinds of possibility distributions of A, namely, the upper and the lower possibility distributions are introduced in this paper. The upper and the lower possibility distributions, denoted as π_u and π_l, respectively, should satisfy the inequality $\pi_u(x) \geq \pi_l(x)$, considering some similarities between our proposed methods and the rough sets.

From the formulation (10.1), it is obvious that the vector r with the highest possibility grade should be closest to the center vector a among all r_i ($i = 1, 2, \cdots,$ m). Thus, the center vector a can be approximately estimated as

$$a = r_{i*}, \tag{10.2}$$

where r_{i*} denotes the vector with grade $h_{i*} = \max_{k=1,2,\cdots,m} h_k$.

The associated possibility grade of r_{i*} is revised to be 1. Taking the transformation $y = r - a$, the possibility distribution with a zero center vector is obtained as:

$$\pi_A(y) = \exp\{-y^T D_A^{-1} y\}.$$

The upper and the lower distributions are used to reflect two kinds of distributions from the upper and the lower directions. In order to determine the matrix π_u in the upper distribution, the following assumptions are given:

1. $\pi_u(y_i) \geq h_i, i = 1, 2, \cdots, m$(the constraint conditions),
2. minimize $\pi_u(y_1) \times \pi_u(y_2) \times \cdots \times \pi_u(y_m)$(the objective function).

Furthermore, the constraint condition can be represented by

$$\pi_u(y_i) \geq h_i \Longleftrightarrow y_i^T D_u^{-1} y_i \leq -\ln h_i.$$

The objective function can be represented by

$$\max \sum_{i=1}^{m} y_i^{\mathrm{T}} D_{\mathrm{u}}^{-1} y_i.$$

Hence, one can get D_{u} by solving the following optimization problem:

$$
\begin{aligned}
\text{(Du)} \quad \max_{D_{\mathrm{u}}} \ & \sum_{i=1}^{m} y_i^{\mathrm{T}} D_{\mathrm{u}}^{-1} y_i \\
\text{s.t.} \ & y_i^{\mathrm{T}} D_{\mathrm{u}}^{-1} y_i \leq -\ln(h_i), \quad i = 1, 2, \cdots, m, \\
& D_{\mathrm{u}} \succ 0,
\end{aligned}
$$

where, $D_{\mathrm{u}} \succ 0$ denotes D_{u} is a definite matrix.

Similarly, in order to determine the matrix π_{l} in the lower distribution, the following assumptions are given:

1. $\pi_{\mathrm{l}}(y_i) \leq h_i, i = 1, 2, \cdots, m$(the constraint condition)
2. maximize $\pi_{\mathrm{l}}(y_1) \times \pi_{\mathrm{l}}(y_2) \times \cdots \times \pi_{\mathrm{l}}(y_m)$(the objective function).

Furthermore, the constraint condition can be represented by:

$$\pi_{\mathrm{l}}(y_i) \leq h_i \iff y_i^{\mathrm{T}} D_{\mathrm{l}}^{-1} y_i \geq -\ln h_i.$$

The objective function can be represented by

$$\min \sum_{i=1}^{m} y_i^{\mathrm{T}} D_{\mathrm{l}}^{-1} y_i.$$

Hence, we can get D_{l} by solving the following optimization problem

$$
\begin{aligned}
\text{(Dl)} \quad \min_{D_{\mathrm{l}}} \ & \sum_{i=1}^{m} y_i^{\mathrm{T}} D_{\mathrm{l}}^{-1} y_i \\
\text{s.t.} \ & y_i^{\mathrm{T}} D_{\mathrm{l}}^{-1} y_i \geq -\ln(h_i), \quad i = 1, 2, \cdots, m, \\
& D_{\mathrm{l}} \succ 0,
\end{aligned}
$$

where, $D_{\mathrm{l}} \succ 0$ denote D_{l} is a definite matrix.

If one solves these two optimization problems separately, it can not be ensured that $\pi_{\mathrm{u}}(y) \geq \pi_{\mathrm{l}}(y)$ holds for an arbitrary y. Tanaka and Guo consider the following model, which integrates (Du) and (Dl) to find out D_{u} and D_{l}, at the same time, adding a constraint condition, i.e., $D_{\mathrm{u}} - D_{\mathrm{l}}$ is a semi-definite matrix. The optimization problem is represented by

$$
\begin{aligned}
\text{(Dul)} \quad \min_{D_{\mathrm{u}}, D_{\mathrm{l}}} \ & \sum_{i=1}^{m} y_i^{\mathrm{T}} D_{\mathrm{u}}^{-1} y_i - \sum_{i=1}^{m} y_i^{\mathrm{T}} D_{\mathrm{l}}^{-1} y_i \\
\text{s.t.} \ & y_i^{\mathrm{T}} D_{\mathrm{u}}^{-1} y_i \leq -\ln(h_i), \quad i = 1, 2, \cdots, m, \\
& y_i^{\mathrm{T}} D_{\mathrm{l}}^{-1} y_i \geq -\ln(h_i), \quad i = 1, 2, \cdots, m, \\
& D_{\mathrm{u}} - D_{\mathrm{l}} \succeq 0, \\
& D_{\mathrm{l}} \succ 0.
\end{aligned}
$$

In this case, $\pi_u(y)$ and $\pi_l(y)$ are similar to the rough set concept because $D_u - D_l \succeq 0$ ensures $\pi_u(y) \geq \pi_l(y)$. It is obvious that (Dul) is a nonlinear optimization problem which is difficult to solve.

In order to solve the problem (Dul) easily, firstly considering a simple linear programming problem without the conditions $D_u - D_l \succeq 0$ and $D_l \succ 0$ in (Dul).

If the obtained matrices D_u and D_l cannot satisfy the conditions $D_u - D_l \succeq 0$ and $D_l \succ 0$, we introduce the following auxiliary conditions to the constraint conditions of the problem, to obtain positive definite matrices D_u and D_l such that $D_u - D_l \succeq 0$ holds by the linear programming.

$$\text{for} \quad i \in E, \quad y_i^T D_u^{-1} y_i \geq \varepsilon, \tag{10.3}$$

$$\text{for all } i \neq j, \quad i, j \in E, \quad y_i^T D_u^{-1} y_j = 0, \tag{10.4}$$

$$\text{for} \quad i \in E, \quad y_i^T (D_l^{-1} - D_u^{-1}) y_i \geq 0, \tag{10.5}$$

$$\text{for all } i \neq j, \quad i, j \in E, \quad y_i^T D_l^{-1} y_j = 0, \tag{10.6}$$

where E is the index set of n selected independent vectors $\{y_1, y_2, \cdots, y_n\}$ among y_i $(i = 1, 2, \cdots, m)$, while considering the condition $m \gg n$, and ε is a small positive number. It should be noted that the center vector $y_{i*} = \mathbf{0}$ is not included in $\{y_1, y_2, \cdots, y_n\}$. The equalities $(10.3) \sim (10.6)$ are called the orthogonal conditions. It is proved afterwards that the constraint conditions can ensure that $D_u \succ 0$, $D_l \succ 0$ and $D_u \succeq D_l$ hold. Thus, the following LP problem is formed.

$$(\text{LP10-1}) \quad \min_{D_u, D_l} \sum_{i=1}^{m} y_i^T D_u^{-1} y_i - \sum_{i=1}^{m} y_i^T D_l^{-1} y_i$$

$$\begin{aligned}
\text{s.t.} \quad & y_i^T D_u^{-1} y_i \leq -\ln(h_i), \quad i = 1, 2, \cdots, m, \\
& y_i^T D_l^{-1} y_i \geq -\ln(h_i), \quad i = 1, 2, \cdots, m, \\
& y_i^T D_u^{-1} y_i \geq \varepsilon, \quad i \in E, \\
& y_i^T D_u^{-1} y_j = 0, \quad i \neq j, \quad i, j \in E, \\
& y_i^T (D_l^{-1} - D_u^{-1}) y_i \geq 0, \quad i \in E, \\
& y_i^T D_l^{-1} y_j = 0, \quad i \neq j, \quad i, j \in E.
\end{aligned}$$

Theorm10.1 The matrices D_u and D_l obtained from the equalities $(10.3) \sim (10.6)$ satisfy the condition $D_u \succ 0$, $D_l \succ 0$ and $D_u - D_l \succeq 0$.
Proof Because $\{y_1, y_2, \cdots, y_n\}$ are independent vectors in the n-dimensional space, an arbitrary vector z can be represented as

$$z = \lambda_1 y_1 + \lambda_2 y_2 + \cdots + \lambda_n y_n,$$

where λ_i is a real number.

Thus, by using (10.3) and (10.4), for $z \neq 0$, we have

$$z^{\mathrm{T}} D_{\mathrm{u}}^{-1} z = (\lambda_1 y_1 + \lambda_2 y_2 + \cdots + \lambda_n y_n)^{\mathrm{T}} D_{\mathrm{u}}^{-1} (\lambda_1 y_1$$
$$+ \lambda_2 y_2 + \cdots + \lambda_n y_n)$$
$$= \sum_{i=1}^{n} \lambda_i^2 y_i^{\mathrm{T}} D_{\mathrm{u}}^{-1} y_i > 0.$$

It means that $D_{\mathrm{u}} \succ 0$.

From (10.5) one can get

$$y_i^{\mathrm{T}} D_{\mathrm{l}}^{-1} y_i \geq y_i^{\mathrm{T}} D_{\mathrm{u}}^{-1} y_i > 0. \tag{10.7}$$

Thus, by using (10.6) and (10.7), one have

$$z^{\mathrm{T}} D_{\mathrm{l}}^{-1} z = (\lambda_1 y_1 + \lambda_2 y_2 + \cdots + \lambda_n y_n)^{\mathrm{T}} D_{\mathrm{l}}^{-1} (\lambda_1 y_1$$
$$+ \lambda_2 y_2 + \cdots + \lambda_n y_n)$$
$$= \sum_{i=1}^{n} \lambda_i^2 y_i^{\mathrm{T}} D_{\mathrm{l}}^{-1} y_i > 0.$$

It means that $D_{\mathrm{l}} \succ 0$. In a similar way, one have

$$z^{\mathrm{T}} (D_{\mathrm{l}}^{-1} - D_{\mathrm{u}}^{-1}) z = (\lambda_1 y_1 + \lambda_2 y_2 + \cdots + \lambda_n y_n)^{\mathrm{T}} (D_{\mathrm{l}}^{-1}$$
$$- D_{\mathrm{u}}^{-1})(\lambda_1 y_1 + \lambda_2 y_2 + \cdots + \lambda_n y_n)$$
$$= \sum_{i=1}^{n} \lambda_i^2 y_i^{\mathrm{T}} (D_{\mathrm{l}}^{-1} - D_{\mathrm{u}}^{-1}) y_i \geq 0.$$

It means that $D_{\mathrm{u}} - D_{\mathrm{l}} \succeq 0$.

Theorem 10.2   An optimal solution of (LP10-1) always exists.

Proof   Let us consider a $n \times n$ matrix K, by which a set of linearly independent vectors $\{y_1, y_2, \cdots, y_n\}$ are transformed into a set of orthonormal vectors $\{z_1, z_2, \cdots, z_n\}$, where $z_i^{\mathrm{T}} = [0, \cdots, 0, 1, 0, \cdots, 0]^{\mathrm{T}}$. Thus,

$$K(y_1, y_2, \cdots, y_n) = I,$$

where I is the identical matrix.

If we take positive definite matrices

$$D_{\mathrm{u}}^{-1} = q_1 K^{\mathrm{T}} K,$$

and

$$D_{\mathrm{l}}^{-1} = q_2 K^{\mathrm{T}} K,$$

where $q_1 \leq q_2$, it is obvious that D_{u} and D_{l} satisfies (10.4) and (10.6), in addition $D_{\mathrm{l}}^{-1} - D_{\mathrm{u}}^{-1} = (q_2 - q_1) K^{\mathrm{T}} K$. Thus, the constraint conditions of (LP5-1) are transferred into

$$q_1 (K y_i)^{\mathrm{T}} (K y_i) \leq -\ln h_i, \quad i = 1, 2, \cdots, m,$$

$$q_2(Ky_i)^{\mathrm{T}}(Ky_i) \leq -\ln h_i, \quad i = 1, 2, \cdots, m,$$

$$q_1(Ky_i)^{\mathrm{T}}(Ky_i) \geq \varepsilon, \quad i \in E,$$

$$(q_2 - q_1)(Ky_i)^{\mathrm{T}}(Ky_i) \geq 0, \quad i \in E,$$

$$q_1(Ky_i)^{\mathrm{T}}(Ky_j) = 0, \quad i \neq j, \quad i \in E,$$

$$q_2(Ky_i)^{\mathrm{T}}(Ky_j) = 0, \quad i \neq j, \quad i \in E. \tag{10.8}$$

If taking

$$q_1 = \min_{i=1,\cdots,m-1, i \neq i^*} \left(-\ln \frac{h_i}{(Ky_i)^{\mathrm{T}}(Ky_j)} \right), \tag{10.9}$$

$$q_2 = \max_{i=1,\cdots,m-1, i \neq i^*} \left(-\ln \frac{h_i}{(Ky_i)^{\mathrm{T}}(Ky_j)} \right), \tag{10.10}$$

and a very small positive value for ε, the obtained $D_{\mathrm{u}}, D_{\mathrm{l}}$ satisfy all of constraint conditions. It means that there is an admissible set in the constraint conditions of (NLP10-1). It should be noted that the center vector $y_{i^*} = \mathbf{0}$ is omitted in the two inequalities of Eq. (10.8), because $(Ky_{i^*})^{\mathrm{T}}(Ky_{i^*}) = -\ln 1 = 0$. Thus, one considers $\{1, 2, \cdots, m-1\}$ without i^* in determining q_1, q_2.

Here, orthogonal conditions are added to constraint conditions that can confine the matrices $D_{\mathrm{u}}, D_{\mathrm{l}}$ to positive definite matrices and $D_{\mathrm{u}} - D_{\mathrm{l}}$ to a semi-positive definite matrix. However, since there are many orthogonal conditions among independent vectors, it is very hard to select appropriate ones.

To cope with this difficulty, Tanaka and Guo use principle component analysis (PCA) to rotate the given data (y_i, h_i) to obtain a positive definite matrix easily. The data can be transformed by linear transformation T. Columns of T are eigenvectors of the matrix $\Sigma = [\sigma_{ij}]$, where

$$\sigma_{ij} = \frac{\displaystyle\sum_{k=1}^{m}(x_{ki} - a_i)(x_{kj} - a_j)h_k}{\displaystyle\sum_{k=1}^{m} h_k}.$$

Without loss of generality, assume that the rank of Σ is n. It should be noted that $T^{\mathrm{T}}T = I$. Using the linear transformation, the data y can be transformed into $\{z = T^{\mathrm{T}}y\}$. Then we have

$$\Pi_A(z) = \exp\{-z^{\mathrm{T}}T^{\mathrm{T}}D_A^{-1}Tz\}.$$

According to the feature of PCA, $T^{\mathrm{T}}D_A^{-1}T$ is assumed to be a diagonal matrix as follows:

$$T^{\mathrm{T}} D_A^{-1} T = C_A = \begin{pmatrix} C_1 & 0 & 0 & \cdots & 0 \\ 0 & C_2 & 0 & \cdots & 0 \\ & & & & \vdots \\ 0 & 0 & C_3 & & \vdots \\ 0 & 0 & 0 & \cdots & C_n \end{pmatrix}.$$

Denote C_A as C_u and C_1 for the upper and the lower possibility distribution cases, respectively and denote c_{uj} and c_{1j} $(j = 1, 2, \cdots, n)$ as the diagonal elements in C_u and C_1, respectively. The integrated model can be rewritten as follows.

(LP10-2) $\displaystyle \min_{C_u, C_1} \sum_{i=1}^{m} z_i^{\mathrm{T}} C_1^{-1} z_i - \sum_{i=1}^{m} z_i^{\mathrm{T}} C_u^{-1} z_i$

$\quad$ s.t. $\quad z_i^{\mathrm{T}} C_1^{-1} z_i \geq -\ln(h_i), \quad i = 1, 2, \cdots, m,$

$\qquad\qquad z_i^{\mathrm{T}} C_u^{-1} z_i \leq -\ln(h_i), \quad i = 1, 2, \cdots, m,$

$\qquad\qquad c_{uj} \geq \varepsilon, \quad j = 1, 2, \cdots, n,$

$\qquad\qquad c_{1j} \geq c_{uj}, \quad j = 1, 2, \cdots, n,$

where, the condition $c_{1j} \geq c_{uj} \geq \varepsilon > 0$, such that the matrix $D_u - D_1$ is semi-definite, the matrices D_u and D_1 are definite. Thus, we have

$$D_u = T C_u^{-1} T^{\mathrm{T}}, \tag{10.11}$$

$$D_1 = T C_1^{-1} T^{\mathrm{T}}. \tag{10.12}$$

This identification procedure is called as the PCA method. It is simpler than the method based on orthogonal conditions.

Theorem10.3 In the linear programming problem (LP10-2), the matrices C_u and C_1 always exist.

Proof Take $C_u = qI$ and $C_1 = pI$ in (LP10-2). Thus, the constraint conditions of (LP5-2) can be represented by

$$\begin{aligned} p z_i^{\mathrm{T}} z_i &\geq -\ln h_i, \quad i = 1, 2, \cdots, m, \\ q z_i^{\mathrm{T}} z_i &\leq -\ln h_i, \quad i = 1, 2, \cdots, m, \\ q &\geq \varepsilon, \\ p &\geq q. \end{aligned} \tag{10.13}$$

If takeing $\varepsilon \leq q$ and

$$p = \max_{i=1,\cdots,m-1, i \neq i^*} \left(-\ln \frac{h_i}{z_i^{\mathrm{T}} z_i} \right),$$

$$q = \min_{i=1,\cdots,m-1, i \neq i^*} \left(-\ln \frac{h_i}{z_i^{\mathrm{T}} z_i} \right),$$

such that (10.13) holds.

Therefore, there is an admissible set in the constraint conditions of (LP10-2). It should be noted that vector $z_{i^*} = \mathbf{0}$ is omitted, because $z_{i^*}^{\mathrm{T}} z_{i^*} = -\ln 1 = 0$ in Eq. (10.13). Thus, we consider $\{1, 2, \cdots, m-1\}$ without $i^* = 0$.

This theorem implies that using PCA method we always can obtain the matrices D_u and D_l in upper and lower possibility distributions.

Assume that the given data $(y_i, h_i)(i = 1, 2, \cdots, m)$ are obtained from an exponential possibility distribution $Y = (0, A^\nabla)_e$ where the center vector is zero. In other words, the following equations hold.

$$\Pi_Y(y_i) = \exp\{-y_i^T A^{\nabla^{-1}} y_i\} = h_i, \quad i = 1, 2, \cdots, m. \tag{10.14}$$

Considering the following optimization problem for finding out the upper possibility matrix A_u and the lower possibility matrix A_l from the above given data.

$$\text{(LP10-3)} \quad \min_{A_u, A_l} J(A_l, A_u) = \sum_{i=1}^{m} y_i^T A_l^{-1} y_i - \sum_{i=1}^{m} y_i^T A_u^{-1} y_i$$
$$\text{s.t.} \quad y_i^T A_l^{-1} y_i \geq -\ln(h_i), \quad i = 1, 2, \cdots, m,$$
$$y_i^T A_u^{-1} y_i \leq -\ln(h_i), \quad i = 1, 2, \cdots, m.$$

Theorem10.4 The optimal solutions of A_u and A_l in (LP10-3) are A^∇.
Proof The optimization problem (LP10-3) can be separated into the following two optimization problems:

$$\text{(LP5-3U)} \quad \max_{A_u} J_1(A_u) = \sum_{i=1}^{m} y_i^T A_u^{-1} y_i$$
$$\text{s.t.} \quad y_i^T A_u^{-1} y_i \leq -\ln(h_i), \quad i = 1, 2, \cdots, m.$$

$$\text{(LP10-3L)} \quad \min_{A_l} J_1(A_l) = \sum_{i=1}^{m} y_i^T A_l^{-1} y_i$$
$$\text{s.t.} \quad y_i^T A_l^{-1} y_i \geq -\ln(h_i), \quad i = 1, 2, \cdots, m.$$

Since data $(y_i, h_i)(i = 1, 2, \cdots, m)$ are obtained from the exponential possibility distribution $Y = (0, A^\nabla)_e$, the data $(y_i, h_i)(i = 1, 2, \cdots, m)$ satisfy (10.14). Therefore, Ar is an admissible solution of (LP10-3U) and (LP10-3L). Assume that there is another matrix A^∇ such as $J_1(A') > J_1(A^\nabla)$ in (LP10-3U). Then, for some i,

$$y_i^T A'^{-1} y_i > y_i^T A^{\nabla^{-1}} y_i = -\ln h_i.$$

which shows that A' is not admissible. Thus, A^∇ is the optimal solution of (LP10-3U). In the same way, we can prove that the optimal solution of (LP10-3L) is also A^∇. Therefore, both A_u and A_l are A^∇.

This theorem means that the methods for determining an exponential possibility distribution can obtain the actual matrix A^∇ if the given data are governed by an exponential possibility distribution with a distribution matrix A^∇. Moreover, the upper and the lower possibility distributions are equal to A^∇.

10.3 Model Formulation

Return of the portfolio $x = (x_1, \cdots, x_n)$ can be represented by $z = r^{\mathrm{T}}x$. Denote the possibility distribution of z by $\pi_Z(z)$.

Because r is governed by a possibility distribution $(a, D_A)_{\mathrm{e}}$, z is a possibility variable Z. According to the extension principle, the possibility distribution of a portfolio return Z is defined by

$$\pi_Z(z) = \max_{\{r|z=r^{\mathrm{T}}x\}} \exp\{-(r-a)^{\mathrm{T}} D_A^{-1}(r-a)\}. \tag{10.15}$$

By solving the above optimization problem, we can get the possibility distribution

$$\pi_Z(z) = \exp\{-(z - a^{\mathrm{T}}x)^2(x^{\mathrm{T}}D_A x)^{-1}\} = (a^{\mathrm{T}}x, x^{\mathrm{T}}D_A x)_{\mathrm{e}},$$

where $a^{\mathrm{T}}x$ is center value, $x^{\mathrm{T}}D_A x$ is the spread of possibility return Z. Given the lower and the upper possibility distributions, the corresponding portfolio selection models are given as follows:

• Portfolio selection model based on upper possibility distributions

$$\text{(QP10-1)} \quad \min_x x^{\mathrm{T}}D_{\mathrm{u}}x$$
$$\text{s.t.} \ a^{\mathrm{T}}x = c,$$
$$\sum_{i=1}^{n} x_i = 1,$$
$$x_i \geq 0, \quad i = 1, 2, \cdots n.$$

• Portfolio selection model based on lower possibility distribution:

$$\text{(QP10-2)} \quad \min_x x^{\mathrm{T}}D_{\mathrm{l}}x$$
$$\text{s.t.} \ a^{\mathrm{T}}x = c,$$
$$\sum_{i=1}^{n} x_i = 1,$$
$$x_i \geq 0, \quad i = 1, \cdots n,$$

where c is the center value of portfolio possibility return given by the investor.

It is straightforward that the above models are quadratic programming problems minimizing the spread of a possibility portfolio return.

Theorem 10.5   The spread of the possibility return based on the lower possibility distribution is not larger than the one based on the upper possibility distribution.

Proof   Suppose that the optimal solutions obtained from the problems (QP10-1) and (QP10-2) are denoted as x_{u}^* and x_{l}^* respectively, with considering the same center value. According to the feature of the upper and lower possibility distributions, i.e. $D_{\mathrm{u}} - D_{\mathrm{l}} \succeq 0$, the following inequality holds.

$$x_u^{*\mathrm{T}} D_u x_u^* \geq x_u^{*\mathrm{T}} D_l x_u^*.$$

Since x_l^* is the optimal solution of (QP5-2), we have

$$x_u^{*\mathrm{T}} D_l x_u^* \geq x_l^{*\mathrm{T}} D_l x_l^*.$$

Thus,

$$x_u^{*\mathrm{T}} D_u x_u^* \geq x_l^{*\mathrm{T}} D_l x_l^*.$$

The nondominated solutions with considering two objective functions, i.e., the spread and the center of a possibility portfolio in the possibility portfolio selection models can form efficient frontiers.

Definition **10.1**   Efficient frontiers from the upper and lower possibility portfolio selection models are called possibility efficient frontier I and possibility efficient frontier II, respectively.

Definition **10.2**   Two spreads of possibility portfolio returns from the upper and lower possibility distributions with the same given center value form an interval. This interval is called a possibility risk interval.

The possibility risk interval is used to reflect the uncertainty in portfolio selection problems.

10.4 Numerical Example

In order to illustrate the proposed approaches, Tanaka and Guo consider an example shown in Table   10.1 and Table 10.2 introduced by Markowitz. In this example, since we can consider that the recent sample is more similar to the like scenario(s) in the future, it is assumed that the possibility grade hi can be obtained as

$$h_i = 0.2 + 0.7(t-1)/17 \quad (t = 1, 2, \cdots, 18).$$

When taking $\varepsilon = 0.001$, by (10.2), we have the center value variable

$$a = (0.154, 0.176, 0.908, 0.715, 0.469, 0.077, 0.112, 0.756, 0.185)^{\mathrm{T}}.$$

Using models (QP10-1) and (QP10-2), they obtained two possibility efficient frontiers shown in Fig 10.1. We can find that the spread of the possibility portfolio return from (QP10-1) is always larger than that from (QP10-2). This fact stems from the concept of the lower and the upper possibility distributions. They can be regarded as two extreme opinions playing a reference role for an investor. The corresponding risk with $c = 0.3$ is an interval value, i.e., $[0.17978, 0.67318]$, which reflect the uncertainty in real investment problems. Table 10.3 and Table 10.4 show the securities selected by the possibility portfolio selection models (QP10-2) and (QP10-1) in the case of $c = 0.3$, respectively. The result shows that the number of the obtained securities from (QP10-2) is more than the one from (QP10-1). It implies that the portfolio from model (48) tends to take more distributive investment than the one from (QP10-1).

Table 10.1. Historical returns of securities 1

	h_i	1 Am.T	2 A.T.T	3 U.S.S	4 G.M.	5 A.T.Sfe
1937	0.2	−0.305	-0.173	−0.318	−0.477	−0.457
1938	0.241	0.513	0.098	0.285	0.714	0.107
1939	0.282	0.055	0.200	−0.047	0.165	−0.424
1940	0.324	−0.126	0.03	0.104	−0.043	−0.189
1941	0.365	−0.280	−0.183	−0.171	−0.277	0.637
1942	0.406	−0.003	0.067	−0.039	0.476	0.865
1943	0.447	0.428	0.300	0.149	0.225	0.313
1944	0.488	0.192	0.103	0.260	0.290	0.637
1945	0.529	0.446	0.216	0.419	0.216	0.373
1946	0.571	−0.088	−0.046	−0.078	−0.272	−0.037
1947	0.612	−0.127	−0.071	0.169	0.144	0.026
1948	0.653	−y0.015	0.056	−0.035	0.107	0.153
1949	0.694	0.305	0.038	0.133	0.321	0.067
1950	0.735	−0.096	0.089	0.732	0.305	0.579
1951	0.776	0.016	0.090	0.021	0.195	0.040
1952	0.818	0.128	0.083	0.131	0.390	0.434
1953	0.859	−0.010	0.035	0.006	−0.072	−0.027
1954	0.9	0.154	0.176	0.908	0.715	0.469

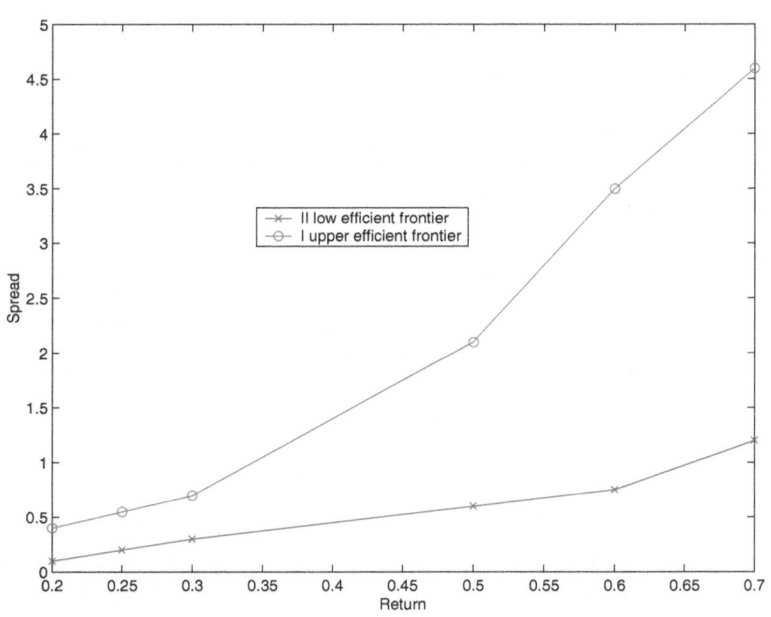

Fig. 10.1. Efficient frontier of possibility model

Table 10.2. Historical returns of securities 2

		6	7	8	9
	h_i	C.C	Bdn	Frstn.	S.S
1937	0.2	−0.065	−0.319	−0.400	−0.435
1938	0.241	0.238	0.076	0.336	0.238
1939	0.282	−0.078	0.381	−0.093	−0.295
1940	0.324	−0.077	−0.051	-y0.090	−0.036
1941	0.365	−0.187	0.087	−0.194	−0.240
1942	0.406	0.156	0.262	0.113	0.126
1943	0.447	0.351	0.341	0.580	0.639
1944	0.488	0.233	0.227	0.473	0.282
1945	0.529	0.349	0.352	0.229	0.578
1946	0.571	−0.209	0.153	−0.126	0.289
1947	0.612	0.355	−0.099	0.009	0.184
1948	0.653	−0.231	0.038	0.000	0.114
1949	0.694	0.246	0.273	0.223	−0.222
1950	0.735	−0.248	0.091	0.650	0.327
1951	0.776	−0.064	0.054	−0.131	0.333
1952	0.818	0.079	0.109	0.175	0.062
1953	0.859	0.067	0.210	−0.084	−0.048
1954	0.9	0.077	0.112	0.756	0.185

Table 10.3. Investment by (QP10-1)

$x_1 = 0.000$	$x_2 = 0.000$	$x_3 = 0.000$
$x_4 = 0.233$	$x_5 = 0.111$	$x_6 = 0.160$
$x_7 = 0.312$	$x_8 = 0.000$	$x_9 = 0.184$

Table 10.4. Investment by (QP10-2)

$x_1 = 0.000$	$x_2 = 0.101$	$x_3 = 0.000$
$x_4 = 0.201$	$x_5 = 0.152$	$x_6 = 0.143$
$x_7 = 0.250$	$x_8 = 0.000$	$x_9 = 0.153$

10.5 Conclusion

In this chapter, we introduce Tanaka and Guo's model (2003). Different from probability distributions for reflecting the statistic characteristics of the past data, possibility distributions are used to characterize human knowledge so that knowledge from multiple experts can be represented by a set of exponential possibility distributions. Based on consistency indices defined by possibility measure of each pair of possibility distributions, a fusion model is proposed to integrate multiple possibility distributions into a new one representing a refined knowledge.

11

Carlsson-Fullér-Majlender's Trapezoidal Possibility Model

11.1 Introduction

In traditional portfolio selection models, uncertainty is regarded as randomness. The probability theory is very useful to deal with observable random events. Though often applied to deal with uncertainty, the probabilistic approaches only partly capture the reality. In reality, although many events are characterized as fuzzy by probabilistic approaches, they are not random events. Carlsson, Fullér and Majlender (2002) have found cases where assignment of probabilities is based on very rough, subjective estimates and then the subsequent calculations are carried out with a precision of two decimal points. The routine use of probabilities is not a good choice is shown. The choice of the utility theory, which builds on a decision maker's relative preferences for artificial lotteries, is a way to anchor portfolio selection in the von NeumannC-Morgenstern axiomatic utility theory. Carlsson, Fullér and Majlender showed that using the utility theory has proved to be problematic: (i) utility measures cannot be validated inter-subjectively, (ii) the consistency of utility measures cannot be validated across events or contexts for the same subject, (iii) utility measures show discontinuities in empirical tests (as shown by Tversky), which should not happen with rational decision makers if the axiomatic foundation is correct, and (iv) utility measures are artificial and intuitive and, thus, hard to use. As the combination of probability assessments with the utility theory has these well-known limitations, Tanaka and Guo (1999) have explored the use of the possibility theory as a substituting conceptual framework. Carlsson, Fullér and Majlender assume that (i) each investor can assign a welfare, or utility, score to competing investment portfolios based on the expected return and risk of the portfolios; and (ii) the rates of return on securities are modeled by possibility distributions rather than probability distributions. They presented an algorithm of complexity $O(n^3)$ for finding an exact optimal solution (in the sense of utility scores) to the n-asset portfolio selection problem under possibility distributions.

11.2 Model Formulation

A fuzzy number A is called trapezoidal with tolerance interval $[a, b]$, left width α and right width β if its membership function takes the following form:

$$A(t) = \begin{cases} 1 - \frac{a-t}{\alpha} & \text{if } a - \alpha \leq t \leq a, \\ 1 & \text{if } a \leq t \leq b, \\ 1 - \frac{t-b}{\beta} & \text{if } a \leq t \leq b + \beta, \\ 0 & \text{otherwise} \end{cases} \qquad (11.1)$$

and we denote $A = (a, b, \alpha, \beta)$. It can easily be shown that

$$[A]^\gamma = [a - (1-\gamma)\alpha, b + (1-\gamma)\beta], \forall \gamma \in [0, 1], \qquad (11.2)$$

where $[A]^\gamma$ denotes the γ-level set of A.

Let $[A]^\gamma = [a_1(\gamma), a_2(\gamma)]$ and $[B]^\gamma = [b_1(\gamma), b_2(\gamma)]$ be fuzzy numbers and let $k \in R$ be a real number. Using the extension principle we can verify the following rules for addition and scalar multiplication of fuzzy numbers:

$$[A + B]^\gamma = [a_1(\gamma) + b_1(\gamma), a_2(\gamma) + b_2(\gamma)], \qquad (11.3)$$

$$[kA]^\gamma = k[A]^\gamma. \qquad (11.4)$$

Carlsson and Fullér (2001) introduced the notation of crisp possibilistic mean value and crisp possibilistic variance of continuous possibility distributions, which are consistent with the extension principle. The crisp possibilistic mean value of A is

$$E(A) = \int_0^1 \gamma(a_1(\gamma) + a_2(\gamma))d\gamma. \qquad (11.5)$$

It is clear that if $A = (a, b, \alpha, \beta)$ is a trapezoidal fuzzy number, then

$$E(A) = \int_0^1 \gamma[a - (1-\gamma)\alpha + b + (1-\gamma)\beta]d\gamma = \frac{a+b}{2} + \frac{\beta - \alpha}{6} \qquad (11.6)$$

In many important cases, it might be easier to estimate the possibility distributions of rates of return on securities, rather than the corresponding probability distributions. Carlsson, Fullér and Majlender suppose that return of the portfolio P is r_P, the expected return is $E(r_P)$, variance of return is $\sigma^2(r_P)$, and the utility function is

$$U(P) = E(r_P) - 0.005 \times A \times \sigma^2(r_P),$$

where A is an index of the investors risk aversion ($A \approx 2.46$ for an average investor in the USA). Consider the following portfolio selection problem with possibility distributions

$$\text{(UP11-1)} \quad \max U \left(\sum_{i=1}^{n} r_i x_i \right) = E \left(\sum_{i=1}^{n} r_i x_i \right)$$

$$-0.005 \times A \times \sigma^2 \times \left(\sum_{i=1}^{n} r_i x_i \right)$$

$$\text{s.t.} \quad \sum_{i=1}^{n} x_i = 1,$$

$$x_i \geq 0, \quad i = 1, 2, \cdots, n,$$

where $r_i = (a_i, b_i, \alpha_i, \beta_i)(i = 1, 2, \cdots, n)$ is a trapezoidal fuzzy number. One can compute easily

$$E \left(\sum_{i=1}^{n} r_i x_i \right) = \sum_{i=1}^{n} \frac{1}{2} \left[a_i + b_i + \frac{\beta_i - \alpha_i}{3} \right] x_i$$

and

$$\sigma^2 \left(\sum_{i=1}^{n} r_i x_i \right) = \left(\sum_{i=1}^{n} \frac{1}{2} \left[b_i - a_i + \frac{1}{3}(\alpha_i + \beta_i) \right] x_i \right)^2$$

$$+ \frac{1}{72} \left[\sum_{i=1}^{n} (\alpha_i + \beta_i) x_i \right]^2 .$$

Introducing the notations:

$$u_i = \frac{1}{2} \left[a_i + b_i + \frac{1}{3}(\beta_i - \alpha_i) \right],$$

$$v_i = \frac{\sqrt{0.005A}}{2} \left[b_i - a_i + \frac{1}{3}(\alpha_i + \beta_i) \right],$$

$$w_i = \frac{\sqrt{0.005A}}{\sqrt{72}} (\alpha_i + \beta_i),$$

Thus, they represent the ith asset by a triplet (v_i, w_i, u_i), where u_i denotes its possibilistic expected value, and $v_i^2 + w_i^2$ its possibilistic variance multiplied by the constant $0.005A$. Assume that there are at least three distinguishable assets, with the assumption that if two assets have the same expected value and variance then they are considered indistinguishable (or identical in the framework of mean-variance analysis). That is, assume that $u_i \neq u_j$ or $v_i^2 + w_i^2 \neq v_j^2 + w_j^2$ for $i \neq j$.

Carlsson, Fullér and Majlender presented the following possibilistic portfolio selection problem

$$\text{(UP11-2)} \quad \max \langle u, x \rangle - \langle v, x \rangle^2 - \langle w, x \rangle^2$$

$$\text{s.t.} \quad \sum_{i=1}^{n} x_i = 1,$$

$$x_i \geq 0, \quad i = 1, 2, \cdots, n.$$

The convex hull of $\{(v_i, w_i, u_i) : i = 1, 2, \cdots, n\}$, denoted by T,

$$T = \text{conv}\{(v_i, w_i, u_i) : i = 1, 2, \cdots, n\}$$
$$= \left\{ \left(\sum_{i=1}^{n} v_i x_i, \sum_{i=1}^{n} w_i x_i, \sum_{i=1}^{n} u_i x_i \right) : \sum_{i=1}^{n} x_i = 1, x_i \geq 0, i = 1, 2, \cdots, n \right\}$$

T is a convex polytope in R^3.

Then (UP11-2) turns into the following three-dimensional non-linear programming problem:

$$(UP11\text{-}3) \quad \max \ -(v_0^2 + w_0^2 - u_0)$$
$$\text{s.t.} \ (v_0, w_0, u_0) \in T,$$

or, equivalently,

$$(UP11\text{-}4) \quad \min \ v_0^2 + w_0^2 - u_0$$
$$\text{s.t.} \ (v_0, w_0, u_0) \in T,$$

where T is a compact and convex subset of R^3, and the implicit function

$$g_c(v_0, w_0) := v_0^2 + w_0^2 - c$$

c is strictly convex for any $c \in R$. This means that any optimal solution to (UP11-4) must be on the boundary of T.

Carlsson, Fullér and Majlender presented an algorithm for finding an optimal solution to (UP11-2) on the boundary of T. Note that, T is a compact and convex polyhedron of R^3 and that any optimal solution to (UP11-4) must be on the boundary of T, which imply that any optimal solution can be obtained as a convex combination of at most 3 extreme points of T. In the algorithm, by lifting the non-negativity conditions for investment proportions one shall calculate: (i) the (exact) solutions to all conceivable 3-asset problems with non-colinear assets, (ii) the (exact) solutions to all conceivable 2-assets problems with distinguishable assets, and (iii) the utility value of each asset. Then one can compare the utility values of all feasible solutions (i.e. solutions with non-negative weights) and portfolios with the highest utility value will be chosen as optimal solutions to the portfolio selection problem (UP11-4). Their algorithm will require $o(n^3)$ steps, where n is the number of available securities.

Consider three assets $(v_i, w_i, u_i)(i = 1, 2, 3)$, which are not colinear: there are not exist $(\alpha_1, \alpha_2, \alpha_3) \in R^3, (\alpha_1, \alpha_2, \alpha_3) \neq 0$, such that

$$\alpha_1 \begin{bmatrix} v_1 \\ w_1 \\ u_1 \end{bmatrix} + \alpha_2 \begin{bmatrix} v_2 \\ w_2 \\ u_2 \end{bmatrix} - (\alpha_1 + \alpha_2) \begin{bmatrix} v_3 \\ w_3 \\ u_3 \end{bmatrix} = 0,$$

Then the 3-asset optimal portfolio selection problem with not-necessarily non-negative weights reads

$$(UP11\text{-}5) \quad \min \ (v_1 x_1 + v_2 x_2 + v_3 x_3)^2 + (v_1 x_1 + v_2 x_2 + v_3 x_3)^2$$
$$- (u_1 x_1 + u_2 x_2 + u_3 x_3)$$
$$\text{s.t.} \ x_1 + x_2 + x_3 = 1.$$

Let

$$L(x, \lambda) = (v_1 x_1 + v_2 x_2 + v_3 x_3)^2 + (v_1 x_1 + v_2 x_2 + v_3 x_3)^2$$
$$-(u_1 x_1 + u_2 x_2 + u_3 x_3) + \lambda(x_1 + x_2 + x_3 - 1). \tag{11.7}$$

$L(x, \lambda)$ is the Lagrange function of the constrained optimization problem (UP11-5). The KuhnCTucker necessity conditions are

$$2v_1(v_1 x_1 + v_2 x_2 + v_3 x_3) + 2w_1(w_1 + w_2 + w_3) - u_1 + \lambda = 0,$$

$$2v_2(v_1 x_1 + v_2 x_2 + v_3 x_3) + 2w_2(w_1 + w_2 + w_3) - u_2 + \lambda = 0,$$

$$2v_3(v_1 x_1 + v_2 x_2 + v_3 x_3) + 2w_3(w_1 + w_2 + w_3) - u_3 + \lambda = 0,$$
$$x_1 + x_2 + x_3 = 1.$$

which leads to the following linear equality system:

$$\begin{bmatrix} q_1^2 + r_1^2 & q_1 q_2 + r_1 r_2 \\ q_1 q_2 + r_1 r_2 & q_2^2 + r_2^2 \end{bmatrix} \begin{bmatrix} x_1 \\ x_2 \end{bmatrix}$$
$$= \begin{bmatrix} 1/2(u_1 - u_3) - q_1 v_3 - r_1 w_3 \\ 1/2(u_2 - u_3) - q_2 v_3 - r_2 w_3 \end{bmatrix}, \tag{11.8}$$

where $q_1 = v_1 - v_3, q_2 = v_2 - v_3, r_1 = w_1 - w_3$ and $r_2 = w_2 - w_3$.

They prove that, if $(v_i, w_i, u_i)(i = 1, 2, 3)$ are not co-linear, then (11.8) has a unique solution. Suppose the solution of (11.8) is not unique, i.e.,

$$\det \begin{bmatrix} q_1^2 + r_1^2 & q_1 q_2 + r_1 r_2 \\ q_1 q_2 + r_1 r_2 & q_2^2 + r_2^2 \end{bmatrix} = 0,$$

i.e.,

$$\det \begin{bmatrix} q_1^2 + r_1^2 & q_1 q_2 + r_1 r_2 \\ q_1 q_2 + r_1 r_2 & q_2^2 + r_2^2 \end{bmatrix}$$
$$= (q_1^2 + r_1^2)(q_2^2 + r_2^2) - (q_1 q_2 + r_1 r_2)^2$$
$$= (q_1 r_2 - q_2 r_1)^2 = \left(\det \begin{bmatrix} q_1 & r_1 \\ q_2 & r_2 \end{bmatrix} \right)^2 = 0.$$

Thus, the rows of $\begin{bmatrix} q_1 & r_1 \\ q_2 & r_2 \end{bmatrix}$ are not linearly independent $\exists (\alpha_1, \alpha_2) \neq \mathbf{0}$, such that

$$\alpha_1 [q_1, r_1] + \alpha_2 [q_2, r_2] = 0$$
$$\Leftrightarrow \alpha_1 [v_1 - v_3, w_1 - w_3] + \alpha_2 [v_2 - v_3, w_2 - w_3] = 0. \tag{11.9}$$

We can find that (11.8) turns into

$$(q_2^2 + r_2^2) \begin{bmatrix} \alpha_2^2 & -\alpha_1 \alpha_2 \\ -\alpha_1 \alpha_2 & \alpha_1^2 \end{bmatrix} \begin{bmatrix} x_1 \\ x_2 \end{bmatrix}$$
$$= \alpha_1 \begin{bmatrix} 1/2\alpha_1(u_1 - u_3) + \alpha_2(q_2 v_3 + r_2 v_3) \\ 1/2\alpha_1(u_2 - u_3) - \alpha_1(q_2 v_3 + r_2 v_3) \end{bmatrix}.$$

Multiplying both sides by $[\alpha_1, \alpha_2]$, we get that u_1, u_2 and u_3 have to satisfy the equation

$$\alpha_1^2 \left[\frac{1}{2}\alpha_1(u_1 - u_3) + \frac{1}{2}\alpha_2(u_2 - u_3) \right] = 0.$$

If $\alpha_1 \neq 0$, then we obtain $\alpha_1(u_1 - u_3) + \alpha_2(u_2 - u_3) = 0$, and from (11.9) it follows that

$$\alpha_1 \begin{bmatrix} v_1 \\ w_1 \\ u_1 \end{bmatrix} + \alpha_2 \begin{bmatrix} v_2 \\ w_2 \\ u_2 \end{bmatrix} - (\alpha_1 + \alpha_2) \begin{bmatrix} v_3 \\ w_3 \\ u_3 \end{bmatrix} = \mathbf{0},$$

i.e., $(v_i, w_i, u_i)(i = 1, 2, 3)$ were colinear.

If $\alpha_1 = 0$, then $\alpha_2 \neq 0$, and from (11.9) we can get $q_2 = r_2 = 0$. Now we find (11.8) turns into

$$\begin{bmatrix} q_1^2 + r_1^2 & 0 \\ 0 & 0 \end{bmatrix} \begin{bmatrix} x_1 \\ x_2 \end{bmatrix} = \begin{bmatrix} 1/2(u_1 - u_3) - q_1 v_3 - r_1 v_3 \\ 1/2(u_2 - u_3) \end{bmatrix}.$$

Multiplying both sides by $[0, 1]$, we obtain

$$\frac{1}{2}(u_2 - u_3) = 0.$$

we can find

$$v_2 - v_3 = w_2 - w_3 = u_2 - u_3 = 0,$$

it means $(v_i, w_i, u_i)(i = 1, 2, 3)$ were colinear.

Using the general inversion formula,

$$\begin{bmatrix} t_1 & t_2 \\ t_3 & t_4 \end{bmatrix}^{-1} = \frac{1}{t_1 t_4 - t_2 t_3} \begin{bmatrix} t_4 & -t_2 \\ -t_3 & t_1 \end{bmatrix},$$

One can find the optimal solution of (11.8)

$$\begin{bmatrix} x_1^* \\ x_2^* \end{bmatrix} = \frac{1}{(q_1 r_2 - q_2 r_1)^2} \times \begin{bmatrix} q_2^2 + r_2^2 & -(q_1 q_2 + r_1 r_2) \\ -(q_1 q_2 + r_1 r_2) & q_1^2 + r_1^2 \end{bmatrix}$$
$$\times \begin{bmatrix} 1/2(u_1 - u_3) - q_1 v_3 - r_1 v_3 \\ 1/2(u_2 - u_3) - q_2 v_3 - r_2 v_3 \end{bmatrix}. \tag{11.10}$$

It is shown that $x^* = (x_1^*, x_2^*, 1 - x_1^* - x_2^*)$ satisfies the Kuhn-Tucker sufficiency condition, i.e. $L''(x, \lambda)$ is a positive definite matrix at $x = x^*$. in the subset defined by

$$\{y = (y_1, y_2, y_3) \in R^3 : y_1 + y_2 + y_3 = 0\}.$$

In fact, from (11.7) one can get

$$M := \tfrac{1}{2} L^{''}(x^*, \lambda)$$

$$= \begin{bmatrix} v_1^2 + w_1^2 & v_1 v_2 + w_1 w_2 & v_1 v_3 + w_1 w_3 \\ v_1 v_2 + w_1 w_2 & v_2^2 + w_2^2 & v_2 v_3 + w_2 w_3 \\ v_1 v_3 + w_1 w_3 & v_2 v_3 + w_2 w_3 & v_3^2 + w_3^2 \end{bmatrix}$$

$$= \begin{bmatrix} v_1 \\ v_2 \\ v_3 \end{bmatrix} \begin{bmatrix} v_1 \\ v_2 \\ v_3 \end{bmatrix}^T + \begin{bmatrix} w_1 \\ w_2 \\ w_3 \end{bmatrix} \begin{bmatrix} w_1 \\ w_2 \\ w_3 \end{bmatrix}^T ,$$

Hence,

$$y^T M y = (v_1 y_1 + v_2 y_2 + v_3 y_3)^2 + (w_1 y_1 + w_2 y_2 + w_3 y_3)^2 \geq 0, \qquad (11.11)$$

holds for any $y \in R^3$. So M is a semi-definite matrix. If $y^T M y = 0$ for some $y = (y_1, y_2, y_3) \neq \mathbf{0}$, $y_1 + y_2 + y_3 = 0$, then from (11.11) one can find

$$v_1 y_1 + v_2 y_2 + v_3 y_3 = 0, \quad w_1 y_1 + w_2 y_2 + w_3 y_3 = 0,$$

and one can get

$$\det \begin{bmatrix} v_1 & v_2 & v_3 \\ w_1 & w_2 & w_3 \\ 1 & 1 & 1 \end{bmatrix} = \det \begin{bmatrix} q_1 & q_2 \\ r_1 & r_2 \end{bmatrix} = \det \begin{bmatrix} q_1 & r_1 \\ q_2 & r_2 \end{bmatrix} = 0,$$

which would lead to a contradiction with the noncolinearity condition. So $L^{''}$ is positive definite. Thus x^* is the unique optimal solution to (UP11-5) and x^* is an optimal solution to (UP11-4)$(n = 3)$ if $x_1^* > 0, x_2^* > 0$ and $x_3^* > 0$,(the Kuhn-Tucker regularity condition). The optimal value of (UP11-5) will be denoted by U_*.

Consider a 2-asset problem with two assets, say (v_1, w_1, u_1) and (v_2, w_2, u_2), such that $(v_1, w_1, u_1) \neq (v_2, w_2, u_2)$. The portfolio selection problem is presented by

(UP11-6) $\min (v_1 x_1 + v_2 x_2)^2 + (v_1 x_1 + v_2 x_2)^2 - (u_1 x_1 + u_2 x_2)$
            s.t. $x_1 + x_2 = 1$.

Let

$$L(x, \lambda) = (v_1 x_1 + v_2 x_2)^2 + (v_1 x_1 + v_2 x_2)^2 - (u_1 x_1 + u_2 x_2) + \lambda(x_1 + x_2 - 1),$$

$L(x, \lambda)$ is the Lagrange function of the constrained optimization problem (UP11-6). The KuhnCTucker necessity conditions are

$$2 v_1 (v_1 x_1 + v_2 x_2) + 2 w_1 (w_1 + w_2) - u_1 + \lambda = 0,$$
$$2 v_2 (v_1 x_1 + v_2 x_2) + 2 w_2 (w_1 + w_2) - u_2 + \lambda = 0,$$
$$x_1 + x_2 = 1.$$

Thus, one can get:

$$[(v_1 - v_2)^2 + (w_1 - w_2)^2]x_1 = \frac{1}{2}(u_1 - u_2) - (v_1 - v_2)v_2 - (w_1 - w_2)w_2. \quad (11.12)$$

If $(v_1 - v_2)^2 + (w_1 - w_2)^2 \neq 0$, one can find

$$x^* = (x_1^*, 1 - x_1^*),$$

where

$$x_1^* = \frac{1}{(v_1-v_2)^2+(w_1-w_2)^2}[\tfrac{1}{2}(u_1 - u_2) \\ -(v_1 - v_2)v_2 - (w_1 - w_2)w_2], \quad (11.13)$$

x^* is the unique solution of (11.12). If $v_1 = v_2$ and $w_1 = w_2$, then from (11.12) one can find $u_1 = u_2$, u2, which would contradict the initial assumption that the two assets are not identical. It can be easily found that $L''(x^*, \lambda)$ is a positive definite matrix in the subset defined by

$$\{y = (y_1, y_2, y_3) \in R^3 : y_1 + y_2 = 0\}.$$

Therefore, x^* is the unique solution of (UP11-6), and if $x_1^* > 0, x_2^* > 0$, x^* is the optimal solution of (UP11-4)$(n = 2)$.

11.3 Algorithm

Carlsson, Fullér and Majlenderan proposed an algorithm for finding an optimal solution to the n-asset possibility portfolio selection problem (UP11-4) is proposed, The algorithm will terminate in $o(n^3)$ steps.

Step 1: Let $c := +\infty$ and $x_c := [0, \cdots, 0]$.

Step 2: Choose three points from the bag $\{(v_i, w_i, u_i) : i = 1, 2, \cdots, n\}$ which have not been considered yet. If there are no such points then go to Step 9, otherwise denote these three points $(v_j, w_j, u_j), (v_k, w_k, u_k)$ and (v_l, w_l, u_l). Let $(v_1, w_1, u_1) := (v_j, w_j, u_j), (v_2, w_2, u_2) := (v_k, w_k, u_k)$ and $(v_3, w_3, u_3) := (v_l, w_l, u_l)$.

Step 3: If

$$\det \begin{bmatrix} q_1 & r_1 \\ q_2 & r_2 \end{bmatrix} = \det \begin{bmatrix} v_1 - v_3 & w_1 - w_3 \\ v_2 - v_3 & w_2 - w_3 \end{bmatrix} = 0,$$

then go to Step 2, otherwise go to Step 4.

Step 4: Using (11.10), compute the two component, x_1^*, x_2^* of the optimal solution of (UP5-6).

Step 5: If $[x_1^*, x_2^*, 1 - x_2^* - x_1^*] > 0$, then go to Step 6, otherwise go to Step 2.

Step 6: If $U_* < c$, then go to Step 7, otherwise go to Step 2.

Step 7: Let $c = U_*$, where U_* is the optimal value of (UP11-6), and let $x_c = [0, \cdots, 0, x_1^*, 0, \cdots, 0, x_2^*, 0, \cdots, 0, x_3^*, 0, \cdots, 0]$, where x_1^* is the jth component of x_c, x_2^* is the kth component of x_c, x_3^* is the lth component of x_c.

Step 8: go to Step 2.

Step 9: Choose two points from the bag $\{(v_i, w_i, u_i) : i = 1, 2, \cdots, n\}$ which have not been considered yet. If there are no such points then go to Step 16, otherwise denote these two points by (v_j, w_j, u_j) and (v_k, w_k, u_k). $(v_1, w_1, u_1) := (v_j, w_j, u_j)$ and $(v_2, w_2, u_2) := (v_k, w_k, u_k)$.

Step 10: If $(v_1 - v_2)^2 + (w_1 - w_2)^2 \neq 0$, then go to Step 9, otherwise go to Step 11.

Step 11: Using (11.13), compute x_1^* the first component of optimal solution of (UP5-6).

Step 12: If $[x_1^*, x_2^*] = [x_1^*, 1 - x_1^*] > 0$, then go to Step 13, otherwise go to Step 9.

Step 13: If $U_* < c$, then go to Step 14, otherwise go to Step 9.

Step 14: Let $c = U_*$, where U_* is the optimal value of (UP5-6), and let $x_c = [0, \cdots, 0, x_1^*, 0, \cdots, 0, x_2^*, 0, \cdots, 0]$, where x_1^* is the jth component of x_c, x_2^* is the kth component of x_c.

Step 15: go to Step 9.

Step 16: Choose a point from the bag $\{(v_i, w_i, u_i) : i = 1, 2, \cdots, n\}$ which has not been considered yet. If there is no such points then go to Step 20, otherwise denote this point by (v_i, w_i, u_i).

Step 17: If $v_i^2 + w_i^2 - u_i < c$, then go to Step 18, otherwise go to Step 16.

Step 18: Let $c = v_i^2 + w_i^2 - u_i$ and let $x_c = [0, \cdots, 0, 1, 0, \cdots, 0]$, where 1 is the ith component of x_c.

Step 19: Go to Step 16.

Step 20: x_c is the optimal solution and $-c$ is the optimal value of the original portfolio selection problem (UP11-4).

11.4 Numerical Example

Carlsson, Fullér and Majlender illustrate the proposed algorithm by a simple example. Consider a 3-asset problem with $A = 2.46$ and with the following possibility distributions:

$$r_1 = (-10.5, 70.0, 4.0, 100.0),$$
$$r_2 = (-8.1, 35.0, 4.4, 54.0),$$
$$r_3 = (-5.0, 28.0, 11.0, 85.0),$$

Hence,

$$(v_1, w_1, u_1) = (6.386, 1.359, 45.750),$$

$$(v_2, w_2, u_2) = (3.469, 0.763, 21.717),$$

$$(v_3, w_3, u_3) = (3.604, 1.255, 23.833).$$

Since

$$\det \begin{bmatrix} q_1 & r_1 \\ q_2 & r_2 \end{bmatrix} = \det \begin{bmatrix} 2.782 & 0.105 \\ -0.135 & -0.491 \end{bmatrix} = -1.352 \neq 0,$$

we can get

$$\begin{bmatrix} x_1^* \\ x_2^* \end{bmatrix} = \frac{1}{-1.352^2} \begin{bmatrix} 0.259 & 0.427 \\ 0.427 & 7.751 \end{bmatrix} \begin{bmatrix} 0.800 \\ 0.044 \end{bmatrix} = \begin{bmatrix} 0.124 \\ 0.373 \end{bmatrix}.$$

Since $[x_1^*, x_2^*, x_3^*] = [0.124, 0.373, 0.503] > 0$, we can get (Step 7)

$U_* = -9.386$ and $x^* = [0.124, 0.373, 0.503]$.

Thus, $x^* = [0.124, 0.373, 0.503]$ is a qualified candidate for being an optimal solution to (UP11-2). Consider all conceivable 2-asset problems (1, 2), (1, 3) and (2, 3), where the numbers stand for the corresponding assets chosen from the bag $\{(v_1, w_1, u_1), (v_2, w_2, u_2), (v_3, w_3, u_3)\}$.

Select (1, 2) since $(v_1 - v_2)^2 + (w_1 - w_2)^2 = 8.864 \neq 0$, we can get

$$U_* = -9.336 \text{ and } [x_1^*, x_2^*] = [0.163, 0.837].$$

Thus, $[0.163, 0.837]$ is a qualified candidate for being an optimal solution to (UP11-2).

Select (1, 3) since $(v_1 - v_3)^2 + (w_1 - w_3)^2 = 7.751 \neq 0$, we can get

$$U_* = -9.352 \text{ and } [x_1^*, x_3^*] = [0.103, 0.897].$$

Thus, $[0.103, 0.897]$ is a qualified candidate for being an optimal solution to (UP11-2).

Select (2, 3) since $(v_2 - v_3)^2 + (w_2 - w_3)^2 = 0.259 \neq 0$, we can get

$$U_* = -9.277 \text{ and } [x_1^*, x_3^*] = [0.171, 0.829].$$

Thus, $[0.171, 0.829]$ is a qualified candidate for being an optimal solution to (UP11-2).

Finally, one can get the utility values of all the three vertexes of the triangle generated by the three assets:

$v_1^2 + w_1^2 - u_1 = -3.122$, $[1, 0, 0]$ is the corresponding feasible solution to (UP11-2);

$v_2^2 + w_2^2 - u_2 = -3.122$, $[0, 1, 0]$ is the corresponding feasible solution to (UP11-2);

$v_3^2 + w_3^2 - u_3 = -3.122$, $[0, 0, 1]$ is the corresponding feasible solution to (UP11-2).

Comparing the utility values of all feasible solutions, one can find that the only solution to the 3-asset problem is $x^* = [0.124, 0.373, 0.503]$ with a utility value of 9.386. The optimal risky portfolio will be preferred to the risk-free investment (by an investor whose degree of risk-aversion is equal to 2.46 if $r_f < 9.386\%$.

11.5 Conclusion

In the chapter, we introduce Carlsson-Fullér-Majlender's trapezoidal possibility Model. They assign a welfare, or utility, score to competing investment portfolios based on the expected return and risk of the portfolios. Assume the rates of return on securities are modeled by trapezoidal possibility distributions. They present an algorithm of complexity $O(n^3)$ for finding an exact optimal solution.

Center Spread Model in Fractional Financial Market

12.1 Estimation of Possibility Distribution by Using Semi-definite Programming

Assume that we have the given data (r_i, h_i) $(i = 1, 2, \cdots, m)$ where $r_i = (r_{1i}, \cdots, r_{ni})^{\mathrm{T}}$ is a vector of returns of n securities i at the ith period and h_i is an associated possibility grade given by expert knowledge to reflect a similarity degree between the future state of stock markets and the state of the ith sample. These grades h_i are assumed to be expressed by a possibility distribution A defined as

$$\pi_A(R) = \exp\{-(R-a)^{\mathrm{T}} D_A^{-1}(R-a)\} = (a, D_A)_{\mathrm{e}}, \qquad (12.1)$$

where a is a center vector and D_A is a symmetric positive definite matrix, denoted as $D_A > 0$

Given the data, the problem is to determine an exponential possibility distribution, i.e., a center vector a and a symmetric positive definite matrix D_A. According to two different viewpoints, two kinds of possibility distributions of A, namely, the upper and the lower possibility distributions, are introduced in this paper. The upper and the lower possibility distributions denoted as π_{u} and π_{l}, respectively, should satisfy the inequality $\pi_{\mathrm{u}}(x) \geq \pi_{\mathrm{l}}(x)$, while considering some similarities between our proposed methods and rough sets.

Similarly, we can get π_{u} and π_{l} by solving the following optimization problem:

$$\text{(Dul)} \quad \min_{D_{\mathrm{u}}, D_{\mathrm{l}}} \sum_{i=1}^{m} y_i^{\mathrm{T}} D_{\mathrm{u}}^{-1} y_i - \sum_{i=1}^{m} y_i^{\mathrm{T}} D_{\mathrm{l}}^{-1} y_i$$

$$\begin{aligned}
\text{s.t.} \quad & y_i^{\mathrm{T}} D_{\mathrm{u}}^{-1} y_i \leq -\ln(h_i), \quad i = 1, 2, \cdots, m, \\
& y_i^{\mathrm{T}} D_{\mathrm{l}}^{-1} y_i \geq -\ln(h_i), \quad i = 1, 2, \cdots, m, \\
& D_{\mathrm{u}} - D_{\mathrm{l}} \succeq 0, \\
& D_{\mathrm{l}} \succ 0.
\end{aligned}$$

It is difficult to solve the above optimization problem. Tanaka and Guo (1999) and Tanaka *et al.* (2000) proposed a method based on orthogonal conditions and a rotation method using the principal component analysis (PCA) to simplify and relax the optimization problem, and then solve the relaxed problem to get an approximate solution. However, the rapidly developing semidefinite programming seems to be a powerful method that can be used to solve the above problem (one can refer to Vandenberghe and Boyd (1996) and Sturm (1999) for details).

One may ask how we can determine the possibility grade h_i properly. The Analytic Hierarchy Process (AHP) [Saaty (1980)] may be a good alternative, since it combines the qualitative and quantitative analysis efficiently.

Based on the semi-definite programming theory, we can get a relaxation of (Dul) as follows

$$(\text{Duls}) \quad \min_{D_{\text{us}}, D_{\text{ls}}} \sum_{i=1}^{m} y_i^{\text{T}} D_{\text{us}} y_i - \sum_{i=1}^{m} y_i^{\text{T}} D_{\text{ls}} y_i$$

$$\begin{aligned}
\text{s.t.} \quad & y_i^{\text{T}} D_{\text{us}} y_i \leq -\ln(h_i), \quad i = 1, 2, \cdots, m, \\
& y_i^{\text{T}} D_{\text{ls}} y_i \geq -\ln(h_i), \quad i = 1, 2, \cdots, m, \\
& D_{\text{us}} - D_{\text{ls}} \succeq 0, \\
& D_{\text{ls}} \succeq 0,
\end{aligned}$$

where, $D_{\text{ls}} \succeq 0$ denotes D_{ls}, a semi-definite matrix.

If we relax the condition that the optimal solution D_{ls}^* and D_{us}^* is positive definite, then $D_{\text{ls}}^{*-1}, D_{\text{us}}^{*-1}$ is the optimal solution of the original problem, let $D_{\text{l}}^* = D_{\text{ls}}^{*-1}$ and $D_{\text{u}}^* = D_{\text{us}}^{*-1}$. Otherwise, if it is semi-definite, then we can add some distribution and fix $D_{\text{ls}}^*, D_{\text{us}}^*$ as follows:

$$D_{\text{ld}} = D_{\text{ls}}^* + \begin{pmatrix} \epsilon_1 & 0 & 0 & \cdots & 0 \\ 0 & \epsilon_2 & 0 & \cdots & 0 \\ \vdots & \vdots & \vdots & & \vdots \\ 0 & 0 & 0 & \cdots & \epsilon_n \end{pmatrix},$$

$$D_{\text{ud}} = D_{\text{us}}^* + \begin{pmatrix} \epsilon_1 & 0 & 0 & \cdots & 0 \\ 0 & \epsilon_2 & 0 & \cdots & 0 \\ \vdots & \vdots & \vdots & & \vdots \\ 0 & 0 & 0 & \cdots & \epsilon_n \end{pmatrix},$$

where $\epsilon_i (i = 1, 2, \cdots, n)$ is small enough to be considered non-negative, which satisfies that inverse matrices of D_{ld} and D_{ud} exist. Then D_{ld}^{-1} and D_{ud}^{-1} are the better approximate solutions of the original problem; let $D_{\text{l}}^* = D_{\text{ld}}^{-1}, D_{\text{u}}^* = D_{\text{ud}}^{-1}$.

12.2 Model Formulation

We assume that an investor allocates his wealth among n risky assets. The investor starts with an existing portfolio and decides how to reconstruct a new portfolio. The investor pays taxes and transaction costs when trading stocks and attempts to maximize the return on the portfolio after paying the taxes and the transaction costs. Return of the portfolio $x = (x_1, \cdots, x_n)$ can be represented by $z = r^{\mathrm{T}}x$. Denote the possibility distribution of z by $\pi_Z(z)$. x_i^0 is the proportion of the security i, $i = 1, 2, \cdots, n$ owned by the investor. k_i is the rate of transaction costs for risky asset i ($i = 1, 2, \cdots, n$). Then the transaction costs of the risky assets can be expressed as

$$C(x) = \sum_{i=1}^{n} C_i(x_i) = \sum_{i=1}^{n} k_i |x_i - x_i^0|.$$

The center value is $\sum_{i=1}^{n} a_i x_i$, the spread is $x^{\mathrm{T}} D_A^* x$. Considering the transaction costs, the most possible value of portfolio return after removing transaction costs is

$$\sum_{i=1}^{n} (a_i x_i - k_i |x_i - x_i^0|).$$

Carlsson and Fullér introduced the notation of crisp possibilistic mean value and crisp possibilistic variance of continuous possibility distributions, which are consistent with the extension principle. The crisp possibilistic mean value of A is

$$E(A) = \int_0^1 \gamma(a_1(\gamma) + a_2(\gamma))d\gamma. \tag{12.2}$$

It is clear that if $A = (a, b, \alpha, \beta)$ is a trapezoidal fuzzy number, then

$$E(A) = \int_0^1 \gamma[a - (1 - \gamma)\alpha + b + (1 - \gamma)\beta]d\gamma = \frac{a + b}{2} + \frac{\beta - \alpha}{6} \tag{12.3}$$

Denote the turnover rate of security j by trapezoidal fuzzy number $\hat{l}_j = (la_j, lb_j, \alpha_j, \beta_j)$. Then the turnover rate of portfolio $x = (x_1, x_2, \cdots, x_n)$ is $\sum_{j=1}^{n} \hat{l}_j x_j$.

By the definition, the crisp possibilistic mean value of the turnover rate of security j is represented as follows:

$$E(\hat{l}_j) = \int_0^1 \gamma[la_j - (1-\gamma)\alpha_j + lb_j + (1-\gamma)\beta_j]d\gamma = \frac{la_j + lb_j}{2} + \frac{\beta_j - \alpha_j}{6}. \tag{12.4}$$

Therefore, the crisp possibilistic mean value of the turnover rate of portfolio $x = (x_1, x_2, \cdots, x_n)$ can be represented as

$$E(\hat{l}(x)) = E(\sum_{j=1}^{n} \hat{l}_j x_j) = \sum_{j=1}^{n} (\frac{la_j + lb_j}{2} + \frac{\beta_j - \alpha_j}{6})x_j. \tag{12.5}$$

In the study, we use the crisp possibilistic mean value of the turnover rate to measure the portfolio liquidity.

The upper possibility distribution portfolio selection model is represented by:

$$(\text{PU12-1}) \quad \max \sum_{i=1}^{n}(a_i x_i - k_i |x_i - x_i^0|)$$

$$\text{s.t.} \quad x^{\mathrm{T}} D_{\mathrm{u}}^* x \le d_0,$$

$$\sum_{j=1}^{n}\left(\frac{la_j + lb_j}{2} + \frac{\beta_j - \alpha_j}{6}\right) x_j \ge E(\hat{l}_0),$$

$$\sum_{i=1}^{n} x_i = 1,$$

$$x_i \ge 0, \quad i = 1, 2, \cdots, n,$$

where d_0 is the tolerance level given by the investor.

The other form of the upper possibility distribution portfolio selection model is represented by

$$(\text{PU12-2}) \quad \min x^{\mathrm{T}} D_{\mathrm{u}}^* x$$

$$\text{s.t.} \quad \sum_{i=1}^{n}(a_i x_i - k_i |x_i - x_i^0|) \ge r_0,$$

$$\sum_{j=1}^{n}\left(\frac{la_j + lb_j}{2} + \frac{\beta_j - \alpha_j}{6}\right) x_j \ge E(\hat{l}_0),$$

$$\sum_{i=1}^{n} x_i = 1,$$

$$x_i \ge 0, \quad i = 1, 2, \cdots, n,$$

where r_0 is the required return level given by the investor.

In the same way, the lower possibility distribution portfolio selection model is represented by:

$$(\text{PL12-1}) \quad \max \sum_{i=1}^{n}(a_i x_i - k_i |x_i - x_i^0|)$$

$$\text{s.t.} \quad x^{\mathrm{T}} D_{\mathrm{l}}^* x \le d_0,$$

$$\sum_{j=1}^{n}\left(\frac{la_j + lb_j}{2} + \frac{\beta_j - \alpha_j}{6}\right) x_j \ge E(\hat{l}_0),$$

$$\sum_{i=1}^{n} x_i = 1,$$

$$x_i \ge 0, \quad i = 1, 2, \cdots, n.$$

The other form of the lower possibility distribution portfolio selection model is represented by:

(PL12-2) $\min x^{\mathrm{T}} D_1^* x$

$$\text{s.t.} \quad \sum_{i=1}^{n} (a_i x_i - k_i |x_i - x_i^0|) \geq r_0,$$

$$\sum_{j=1}^{n} \left(\frac{la_j + lb_j}{2} + \frac{\beta_j - \alpha_j}{6} \right) x_j \geq E(\hat{l}_0),$$

$$\sum_{i=1}^{n} x_i = 1,$$

$$x_i \geq 0, \quad i = 1, 2, \cdots, n.$$

Introducing x_{n+1}, let

$$\sum_{i=1}^{n} k_i |x_i - x_i^0| \leq x_{n+1},$$

$$d_i^+ = \frac{|x_i - x_i^0| + (x_i - x_i^0)}{2}, \quad d_i^- = \frac{|x_i - x_i^0| - (x_i - x_i^0)}{2}.$$

then (PU12-1) can be transformed into

(PU12-3) $\max \sum_{i=1}^{n} a_i x_i - x_{n+1}$

$$\text{s.t.} \quad x^{\mathrm{T}} D_u^* x \leq d_0,$$

$$\sum_{j=1}^{n} \left(\frac{la_j + lb_j}{2} + \frac{\beta_j - \alpha_j}{6} \right) x_j \geq E(\hat{l}_0),$$

$$\sum_{i=1}^{n} k_i (d_i^+ + d_i^-) \leq x_{n+1},$$

$$d_i^+ - d_i^- = x_i - x_i^0, \ i = 1, 2, \cdots, n,$$

$$\sum_{i=1}^{n} x_i = 1$$

$$x_i, d_i^+, d_i^- \geq 0, \ i = 1, 2, \cdots, n.$$

Similarly, (PU12-2) can be transformed into

(PU12-4) $\min x^{\mathrm{T}} D_u^* x$

$$\text{s.t.} \quad \sum_{i=1}^{n} [a_i x_i - k_i (d_i^+ + d_i^-)] \geq r_0,$$

$$\sum_{j=1}^{n} \left(\frac{la_j + lb_j}{2} + \frac{\beta_j - \alpha_j}{6} \right) x_j \geq E(\hat{l}_0),$$

$$d_i^+ - d_i^- = x_i - x_i^0, \ i = 1, 2, \cdots, n,$$

$$\sum_{i=1}^{n} x_i = 1$$

$$x_i, d_i^+, d_i^- \geq 0, \ i = 1, 2, \cdots, n.$$

(PL12-1) can be transformed into

$$(PL12\text{-}3) \quad \max \sum_{i=1}^{n} a_i x_i - x_{n+1}$$

$$\text{s.t.} \quad x^T D_1^* x \leq d_0,$$

$$\sum_{j=1}^{n} \left(\frac{la_j + lb_j}{2} + \frac{\beta_j - \alpha_j}{6} \right) x_j \geq E(\hat{l}_0),$$

$$\sum_{j=1}^{n} k_i(d_i^+ + d_i^-) \leq x_{n+1},$$

$$d_i^+ - d_i^- = x_i - x_i^0, \quad i = 1, 2, \cdots, n,$$

$$\sum_{i=1}^{n} x_i = 1$$

$$x_i, d_i^+, d_i^- \geq 0, \quad i = 1, 2, \cdots, n.$$

(PL12-2) can be transformed into

$$(PL12\text{-}4) \quad \min x^T D_1^* x$$

$$\text{s.t.} \quad \sum_{i=1}^{n} [a_i x_i - k_i(d_i^+ + d_i^-)] \geq r_0,$$

$$\sum_{j=1}^{n} \left(\frac{la_j + lb_j}{2} + \frac{\beta_j - \alpha_j}{6} \right) x_j \geq E(\hat{l}_0),$$

$$d_i^+ - d_i^- = x_i - x_i^0, \quad i = 1, 2, \cdots, n,$$

$$\sum_{i=1}^{n} x_i = 1,$$

$$x_i, d_i^+, d_i^- \geq 0, \quad i = 1, 2, \cdots, n.$$

Both (PL12-3) and (PL12-4) can be used to formulate the efficient frontier of the lower possibility center-spread portfolio selection.

12.3 Numerical Example

In this section, we give an example to illustrate the model for portfolio selection proposed in this chapter. We suppose that an investor wants to choose twelve stocks and a kind of risk-less asset from the Shanghai Stock Exchange for his investment.

The names of the twelve stocks are given in Table 12.1.

In this example, since we can consider that the recent sample is more similar to the future state, it is assumed that the possibility grade hi can be obtained as

$$h_i = 0.2 + 0.7(t - 1)/17 \quad (t = 1, 2, \cdots, 18).$$

Table 12.1. Name of Stocks

Handan Gangtie	Qilu Shihua	Shanghai Jichang
Wukuang Fazhan	Gezhouba	Jiangnan Zhonggong
Guangzhou Konggu	Qinghua Tongfang	Shanghai Jiche
Dongfang Hangkong	Dongfang Jituan	Diyibaihuo

Using the semi-definite programming method, we estimate D_u and D_l. By solving (PL12-4), we get some optimal portfolios in Table 12.2. The detailed investment strategies are listed in Table 12.4. Based on table 12.2, we can describe the lower possibility center spread efficient frontier in Fig 12.1.

Table 12.2. Risk, return and liquidity of the lower possibility portfolio

	1	2	3
return	0.001	0.003	0.004
risk	0.300 7	0.371 7	0.446 4
liquidity	0.056 6	0.047 0	0.047 5
	4	5	6
return	0.005	0.006	0.007
risk	0.547 2	0.672 4	0.832 9
liquidity	0.047 4	0.047 2	0.046 2

By solving (PU12-4), we get some optimal portfolios in Table 12.3. The detailed investment strategies are listed in Table 12.5. Based on table 12.3, we can describe the upper possibility center spread efficient frontier in Fig 12.2.

Table 12.3. Risk, return and liquidity of the upper possibility portfolio

	1	2	3
return	0.001	0.003	0.004
risk	0.702 5	0.712 1	0.843 3
liquidity	0.049 9	0.047 8	0.045 6
	4	5	6
return	0.005	0.006	0.007
risk	1.129 5	1.622 2	2.896 7
liquidity	0.044 5	0.043 2	0.042 7

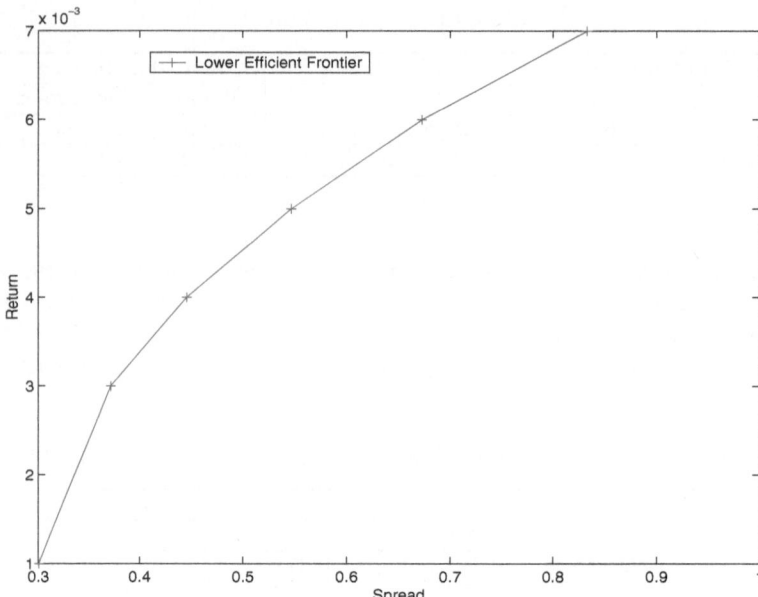

Fig. 12.1. The efficient frontier of lower possibility center spread

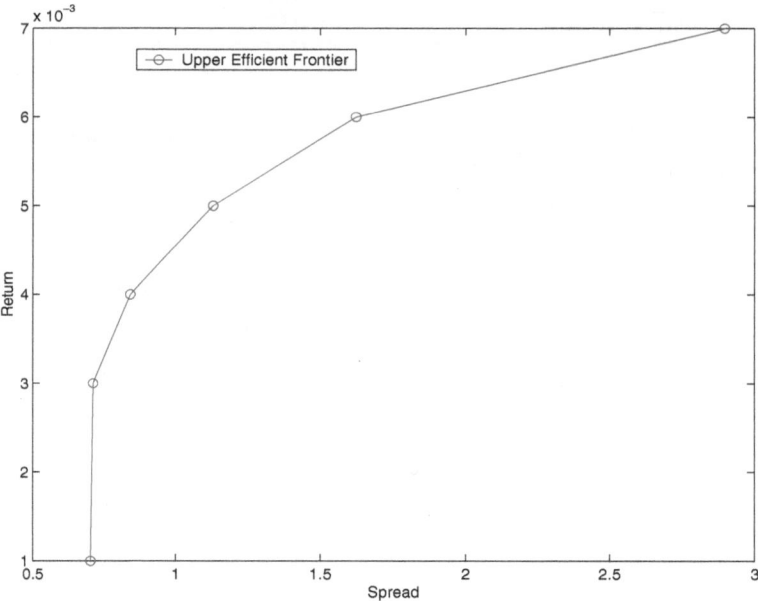

Fig. 12.2. The efficient frontier of upper possibility center spread

Table 12.4. The detailed investment strategies of lower possibility portfolio

Stock	1	2	3
Handan Gangtie	0.560 8	0.317 8	0.203 6
Qilu Shihua	0.000 0	0.000 0	0.000 0
Shanghai Jichang	0.339 3	0.077 8	0.138 6
Wukuang Fazhan	0.000 2	0.000 1	0.003 4
Gezhouba	0.123 7	0.043 1	0.000 0
Guangzhou Konggu	0.000 0	0.000 0	0.000 0
Qinghua Tongfang	0.000 0	0.000 0	0.000 0
Shanghai Qiche	0.315 3	0.478 4	0.518 9
Dongfang Hangkong	0.000 0	0.000 0	0.000 0
Dongfang Jituan	0.000 0	0.082 8	0.138 8
Diyibaihuo	0.000 0	0.000 0	0.000 0
Stock	4	5	6
Handan Gangtie	0.082 0	0.000 1	0.000 0
Qilu Shihua	0.000 0	0.000 0	0.000 0
Shanghai Jichang	0.188 1	0.221 4	0.242 6
Wukuang Fazhan	0.000 0	0.041 1	0.129 7
Gezhouba	0.000 0	0.000 0	0.000 0
Jiangnan Zhonggong	0.000 0	0.000 0	0.000 0
Guangzhou Konggu	0.000 0	0.000 0	0.000 0
Qinghua Tongfang	0.000 0	0.000 0	0.000 0
Shanghai Qiche	0.530 3	0.493 3	0.338 5
Dongfang Hangkong	0.032 0	0.089 0	0.209 8
Dongfang Jituan	0.164 2	0.155 1	0.079 4
Diyibaihuo	0.000 0	0.000 0	0.000 0

12.4 Conclusion

In the chapter, based on the semi-definite programming theory, we present an approach to estimate the possibility distribution of return of security. Then we propose a quadratic programming model for the portfolio selection problem in fractional financial markets. An example is given to illustrate the proposed portfolio selection model.

Table 12.5. The detailed investment strategies of upper possibility portfolio

Stock	1	2	3
Handan Gangtie	0.162 9	0.422 8	0.097 8
Qilu Shihua	0.133 0	0.000 0	0.052 4
Shanghai Jichang	0.095 0	0.016 7	0.136 7
Wukuang Fazhan	0.000 0	0.000 1	0.000 0
Gezhouba	0.129 0	0.103 4	0.105 1
Jiangnan Zhonggong	0.129 0	0.000 0	0.208 7
Guangzhou Konggu	0.089 7	0.000 0	0.087 5
Qinghua Tongfang	0.000 0	0.000 0	0.000 0
Shanghai Qiche	0.084 4	0.434 8	0.109 1
Dongfang Hangkong	0.069 5	0.000 0	0.124 6
Dongfang Jituan	0.000 8	0.022 2	0.000 0
Diyibaihuo	0.106 7	0.000 0	0.078 3
Stock	4	5	6
Handan Gangtie	0.046 8	0.000 0	0.000 0
Qilu Shihua	0.000 0	0.000 0	0.000 0
Shanghai Jichang	0.168 4	0.210 4	0.295 8
Wukuang Fazhan	0.000 0	0.000 0	0.000 0
Gezhouba	0.084 6	0.032 8	0.000 0
Jiangnan Zhonggong	0.267 9	0.334 6	0.225 0
Guangzhou Konggu	0.086 1	0.081 9	0.000 0
Qinghua Tongfang	0.000 0	0.000 0	0.000 0
Shanghai Qiche	0.124 5	0.111 6	0.000 0
Dongfang Hangkong	0.166 7	0.226 5	0.483 2
Dongfang Jituan	0.000 0	0.000 0	0.000 0
Diyibaihuo	0.055 0	0.002 2	0.000 0

Part V

Fuzzy Passive Portfolio Selection Models

13

Fuzzy Index Tracking Portfolio Selection Model

13.1 Introduction

In financial markets, investment strategies can be divided into two classes: passive investment strategies and active investment strategies. Investors adopting active investment strategies trade in securities actively, so that they can find profit opportunities on a running basis. Active investors take it for granted that they can beat markets continuously. Investors who adopt passive investment strategies consider that the securities market is efficient. Therefore, they cannot go beyond the average return level of the market continuously. Index tracking investment is a kind of passive investment strategy, i.e., investors purchase all or some securities which are contained in a securities market index and construct an index tracking portfolio. The securities market index is considered as a benchmark. The investors want to obtain a return similar to that of the benchmark, through index tracking investment.

Roll used the sum of the squared deviations of returns on an index replicating portfolio as the tracking errors and proposed a mean variance index tracking portfolio selection model. Clarke, Krase and Statman defined a linear tracking error, which is the absolute deviation between the managed portfolio return and the benchmark portfolio return. Based on the linear objective function, in which absolute deviations between portfolio and benchmark returns are used, Rudolf, Wolter and Zimmermann proposed four alternative definitions of a tracking error. Furthermore, they gave four linear optimization models for index tracking portfolio selection problem. Consiglio and Zenios and Worzel, Vassiadou-Zeniou and Zenios studied the tracking of indices of fixed-income securities problem. In this chapter, we will use the excess return and the linear tracking error as objective functions and propose a bi-objective programming model for the index tracking portfolio selection problem. Furthermore, we use fuzzy numbers to describe investors' vague aspiration levels for the excess return and the tracking error and propose a fuzzy index tracking portfolio selection model.

The chapter is organized as follows. In Section 2, we present a bi-objective programming model for the index tracking portfolio selection problem. In Section 3, regarding investors' vague aspiration levels for the excess return and linear tracking error as fuzzy numbers, we propose a fuzzy index tracking portfolio selection model. In Section 4, a numerical example is given to illustrate the behavior of the proposed fuzzy index tracking portfolio selection model. Some concluding remarks are given in Section 5.

13.2 Bi-objective Programming Model for Index Tracking Portfolio Selection

We assume that an investor wants to construct a portfolio which is required to track a securities market index. The investor allocates his/her wealth among n risky securities which are component stocks contained in the securities market index. We introduce some notations as follows.

r_{it}: the observed return of security i $(i = 1, 2, \cdots, n)$ at time t $(t = 1, 2, \cdots, T)$;

x_i: the proportion of the total amount of money devoted to security i $(i = 1, 2, \cdots, n)$;

I_t: the observed securities market index return at time t $(t = 1, 2, \cdots, T)$.

Let $x = (x_1, x_2, \cdots, x_n)$. Then the return of portfolio x at time t $(t = 1, 2, \cdots, T)$ is given by

$$R_t(x) = \sum_{i=1}^{n} r_{it} x_i.$$

An excess return is the return of index tracking portfolio x above the return on the index. The excess return of portfolio x at time t $(t = 1, 2, \cdots, T)$ is given by

$$E_t(x) = R_t(x) - I_t.$$

The expected excess return of index tracking portfolio x is given by

$$E(x) = \sum_{t=1}^{T} \frac{1}{T} \left(R_t(x) - I_t \right).$$

Roll used the sum of squared deviations between the portfolio and benchmark returns to measure the tracking error of the index tracking problem. Rudolf, Wolter and Zimmermann used linear deviations instead of squared deviations, to give four definitions of the linear tracking errors. We adopt the tracking error based on the mean absolute downside deviations, to formulate the index tracking portfolio selection model in this paper. The tracking error based on the mean absolute downside deviations can be expressed as

$$T_{DMAD}(x) = \sum_{t=1}^{T} \frac{1}{T} | \min\{0, R_t(x) - I_t\}|.$$

Generally, in the index tracking portfolio selection problem, the tracking error and the excess return are two important factors considered by investors. An investor tries to maximize the expected excess return. At the same time, the investor hopes that the return of portfolio equals the return of the index approximately, to some extent, in the investment horizon. Hence, the expected excess return and the tracking error can be considered as two objective functions of the index tracking portfolio selection problem.

In many financial markets, short selling of securities is not allowed. So we add the following constraints:

$$x_1, x_2, \cdots, x_n \geq 0, \ i = 1, 2, \cdots, n.$$

We assume that the investor pursues maximization of the excess return of portfolio and to minimize the tracking error, under the 'no short selling' constraint. The index tracking portfolio selection problem can be formally stated as the following bi-objective programming problem:

(BP) max $E(x)$
    min $T_{DMAD}(x)$
    s.t. $\sum_{i=1}^{n} x_i = 1,$
    $x_1, x_2, \cdots, x_n \geq 0, \ i = 1, 2, \cdots, n.$

The problem (BP) can be reformulated as a bi-objective linear programming problem by using the following technique. Note that

$$\left| \min\{0, a\} \right| = \frac{1}{2}|a| - \frac{1}{2}a$$

for any real number a. Thus, by introducing auxiliary variables $b_t^+, b_t^-, t = 1, 2, \cdots, T$ such that

$$b_t^+ + b_t^- = \frac{|R_t(x) - I_t|}{2},$$

$$b_t^+ - b_t^- = \frac{R_t(x) - I_t}{2}, \tag{13.1}$$

$$b_t^+ \geq 0, \ b_t^- \geq 0, \ t = 1, 2, \cdots, T, \tag{13.2}$$

we may write

$$T_{DMAD}(x) = \sum_{t=1}^{T} \frac{2b_t^-}{T}.$$

Hence, we may rewrite problem (BP) as the following bi-objective linear programming problem:

(BLP) max $E(x)$

$$\min \sum_{t=1}^{T} \frac{2b_t^-}{T}$$

s.t.   (13.1), (13.2) and all constraints of (BP).

Thus the investor may get the index tracking investment strategies by computing efficient solutions of (BLP). One can use one of the existing algorithms of multiple objective linear programming to solve it efficiently.

13.3 Fuzzy Index Tracking Portfolio Selection Model

As in case of other investments, knowledge and experience of experts are important, for the investor to decide his/her levels of aspiration for the expected excess return and the tracking error of index tracking portfolio also. Watada employed a non-linear S shape membership function, to express aspiration levels of expected return and of risk which the investor would expect, and proposed a fuzzy active portfolio selection model. The S shape membership function is given by:

$$f(x) = \frac{1}{1 + \exp(-\alpha x)}.$$

In the bi-objective programming model of index tracking portfolio selection proposed in Section 2, the two objectives, the expected excess return and the tracking error, are considered. Since the expected excess return and the tracking error are vague and uncertain, we use the non-linear S shape membership functions proposed by Watada to express the aspiration levels of the expected excess return and the tracking error.

The membership function of the expected excess return is given by

$$\mu_E(x) = \frac{1}{1 + \exp\left(-\alpha_E\left(E(x) - E_M\right)\right)},$$

where E_M is the mid-point where the membership function value is 0.5 and α_E can be given by the investor based on his/her own degree of satisfaction for the expected excess return. Figure 13.1 shows the membership function of the goal for the expected excess return.

The membership function of the tracking error is given by

$$\mu_T(x) = \frac{1}{1 + \exp(\alpha_T(T_{DMAD}(x) - T_M))},$$

where T_M is the mid-point where the membership function value is 0.5 and α_T can be given by the investor based on his/her own degree of satisfaction regarding the level of tracking error. Figure 13.2 shows the membership function of the goal for the tracking error.

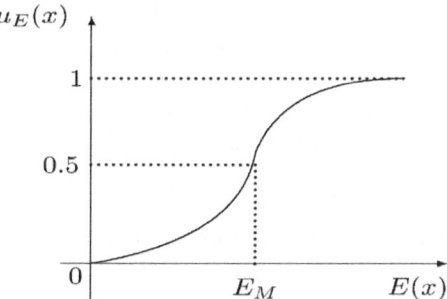

Fig. 13.1. Membership function of the goal for expected excess return

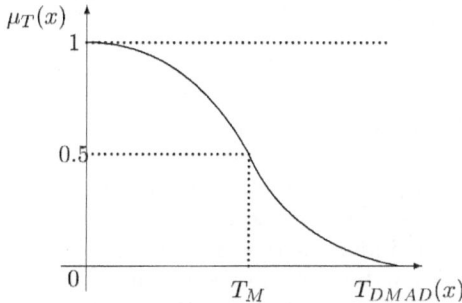

Fig. 13.2. Membership function of the goal for tracking error

Remark1: α_E and α_T determine the shapes of membership functions $\mu_E(x)$ and $\mu_T(x)$ respectively, where $\alpha_E > 0$ and $\alpha_T > 0$. The larger the parameters α_E and α_T get, the lower is their vagueness.

According to Bellman and Zadeh's maximization principle, we can define

$$\lambda = \min \{\mu_E(x), \mu_T(x)\}.$$

The fuzzy index tracking portfolio selection problem can be formulated as follows:

$$\text{(FP)} \max \lambda$$
$$\text{s.t.} \quad \mu_E(x) \geq \lambda,$$
$$\mu_T(x) \geq \lambda,$$
$$\text{and all constraints of (BLP).}$$

Let $\eta = \log\frac{1}{1-\lambda}$, then $\lambda = \frac{1}{1+\exp(-\eta)}$. The logistic function is monotonously increasing, so maximizing λ makes η maximize. Therefore, the above problem can be transformed into an equivalent problem as follows:

$$\text{(FLP)} \max \eta$$
$$\text{s.t.} \quad \alpha_E \left(E(x) - E_M\right) - \eta \geq 0,$$
$$\alpha_T \left(T_{DMAD}(x) - T_M\right) + \eta \leq 0,$$
$$\text{and all constraints of (BLP),}$$

where α_E and α_T are parameters which can be given by the investor, based on his/her own degree of satisfaction regarding the expected excess return and the tracking error.

(FLP) is a standard linear programming problem. One can use one of several algorithms of linear programming to solve it efficiently; for example, the simplex method.

Remark2: The non-linear S shape membership functions of the two factors may change their shape according to parameters α_E and α_T. By selecting the values of these parameters, the aspiration levels of the two factors may be described accurately. On the other hand, different parameter values may reflect different aspiration levels. Therefore, it is convenient for different investors to formulate investment strategies by using the proposed fuzzy index tracking portfolio selection model.

13.4 Numerical Example

In this section, we will give a numerical example to illustrate the proposed fuzzy index tracking portfolio selection model. We suppose that the investor considers Shanghai 180 index as the tracking goal. We choose thirty component stocks from Shanghai 180 index as the risky securities. We collect historical data of the thirty stocks and Shanghai 180 index from January, 1999 to December, 2002. The data can be downloaded from the web-site www.stockstar.com. We use one month as a period to get the historical rates of returns of forty eight periods.

The values of the parameters α_E, α_T, E_M and T_M can be given by the investor according to his/her aspiration levels for the expected excess return and the tracking error. In the example, we assume that $\alpha_E = 500$, $\alpha_T = 1000$, $E_M = 0.010$ and $T_M = 0.009$. Using the historical data, we get an index tracking portfolio selection strategy by solving (FLP). All computations were carried out on a WINDOWS PC using the LINDO solver. Table 13.1 shows the expected excess return and tracking errors of the portfolio, obtained by solving (FLP). Table 13.2 shows the investment ratio of the obtained fuzzy index tracking portfolio. Figure 13.3 shows the deviations between the returns of the obtained index tracking portfolio and the returns on the benchmark Shanghai 180 index for each month from January, 1999 to March, 2003. From Figure 13.3, we can find that the fuzzy index portfolio obtained by solving (FLP) tracks Shanghai 180 index efficiently.

13.5 Conclusion

Regarding the expected excess return and the tracking error as two objective functions, we have proposed a bi-objective programming model for the index tracking portfolio selection problem. Furthermore, investors' vague aspiration

Table 13.1. Membership grade λ, obtained expected excess return and obtained tracking error

λ	η	excess return	tracking error
0.9431	2.8095	0.0152	0.0062

Table 13.2. Investment ratio of the obtained fuzzy index tracking portfolio

Stock	1	2	3	4	5	6	7	8	9	10
Ratio	0.0000	0.0000	0.0620	0.0254	0.0000	0.0408	0.0180	0.1389	0.0324	0.0082
Stock	11	12	13	14	15	16	17	18	19	20
Ratio	0.1440	0.1488	0.0130	0.0000	0.0000	0.0000	0.1889	0.0000	0.0000	0.0000
Stock	21	22	23	24	25	26	27	28	29	30
Ratio	0.0276	0.0000	0.0000	0.0124	0.1001	0.0000	0.0395	0.0000	0.0000	0.0000

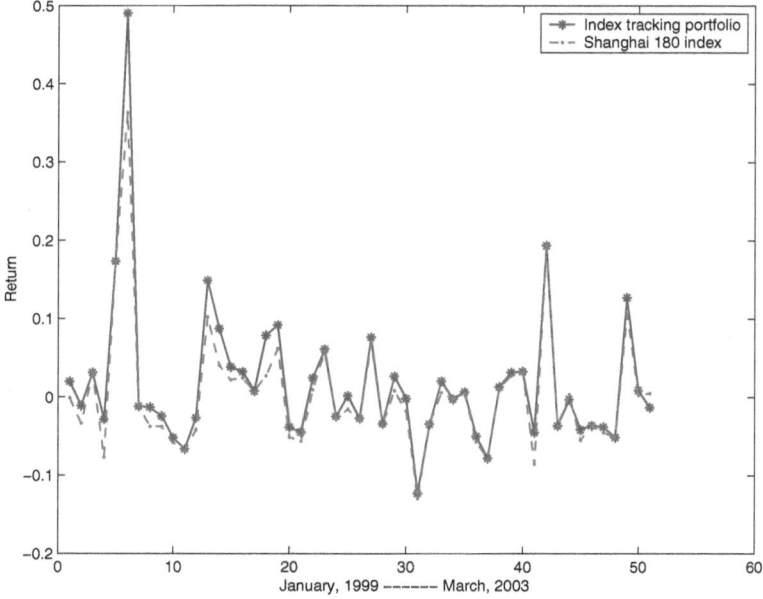

Fig. 13.3. The deviations between the returns of the obtained index tracking portfolio and the returns on the benchmark Shanghai 180 index

levels for the excess return and the tracking error are considered as fuzzy numbers. Based on the fuzzy decision theory, we have proposed a fuzzy index tracking portfolio selection model. The computation results of the example show that the proposed model can generate a favorite index tracking portfolio strategy according to the investor's satisfaction degree expectations.

References

1. Alefeld, G. and Herzberger, J., *Introduction to Interval Computations*, New York: Academic Press, 1983
2. Admati, A. R. and Pfleiderer, P., Does it all add up? Benchmarking and the compensation of active portfolio managers, *Journal of Business*, 1997, Vol. 70, 323–350
3. Arnott, R. D. and Wagner, W. H., The measurement and control of trading costs, *Financial Analysts Journal*, 1990, Vol. 46, 73–80
4. Atknson, C. and Al-ali, B., On an investment-consumption model with transaction cost: an asymptotic analysis, *Applied Mathematical Finance*, 1997, Vol. 4, 109–113
5. Avriel, M., *Nonlinear Programming: Analysis and Methods*, New Jersey: Prentice Hall, 1976
6. Bazaraa, M. S., Sherali, H. D. and Shetty, C. M., *Nonlinear Programming: Theory and Algorithms*, 2nd Edition, New York: John Wiley & Sons, 1993
7. Bell, R. and Cover, T. M., Game-theoretic optimal portfolio, *Management Science*, 1988, Vol. 34, 724–733
8. Bellman, R. and Zadeh, L. A., Decision making in a fuzzy environment, *Management Science*, 1970, Vol. 17, 141–164
9. Bitran, G. R., Linear multiple objective problems with interval coefficients, *Management Science*, 1980, Vol. 26, 694–706
10. Black F., Capital market equilibrium with restricted borrowing, *Journal of Business*, 1972, Vol. 45, 444–455
11. Black, F. and Sholes, M., The pricing of options and corporate liabilities, *Journal of Political Economy*, 1973, Vol. 81, 637–659
12. Brennan, M., Taxes, market valuation and corporate financial policy, *National Tax Journal*, 1970, Vol. 23, 417–427
13. Brennan, M. J., The optimal number of securities in a risky asset portfolio when there are fixed costs of transaction: theory and some empirical results, *Journal of Financial Quantitative Analysis*, 1993, Vol. 10, 487–512
14. Cai, X. Q., Teo, K. L., Yang, X. Q. and Zhou, X. Y., Under a minimax rule portfolio optimization, *Management Science*, 2000, Vol. 46, 957–972
15. Carlsson, C. and Fullér, R., On possibilistic mean value and variance of fuzzy numbers, *Fuzzy Sets and Systems*, 2001, Vol. 122, 315–326

16. Carlsson, C., Fullér, R. and Majlender, P., A possibilistic approach to selecting portfolios with highest utility score, *Fuzzy Sets and Systems*, 2002, Vol. 131, 13–21

17. Campbell, J. and Viceira, L. M., Consumption and portfolio decisions when expected returns are time varying, *Quarterly Journal of Economics*, 1999, Vol. 114, 433–495

18. Chanas, S. and Kuchta, D., Multiobjective programming in optimization of interval objective functions-a generalized approach, *European Journal of Operational Research*, 1996, Vol. 94, 594–598

19. Changkong, V. and Haimes, Y., *Multiobjective Decision Making Theory and Methodology*, Amsterdam: North-Holland, 1983

20. Chao, X., Wang, S. Y. and Yu, M., Multi-period portfolio selection problem with risk control, to appear in *Annals of Operations Research*, 2004

21. Chen, A. H. Y., Jen, F. C. and Zionts, S., The optimal portfolio revision policy, *Journal of Business*, 1971, Vol. 44, 51–61

22. Cheng, S. W., Liu, Y. H. and Wang, S. Y., Progress in risk management, *Advanced Modeling and Optimization*, 2004, Vol. 6, 1–20

23. Chen, G.H., Chen, S., Fang, Y., Wang, S.Y., A possibility mean VaR model for portfolio selection, *AMO-Advanced Modeling and Optimization*, Vol.8(1), 2006, 99–107

24. Chen, S. J. and Hwang, C. L., *Fuzzy Multiple Attribute Decision Making: Methods and Application*, New York: Springer-Verlag, 1992

25. Chunhachinda, P., Dandapani, K., Hamid, S. and Prakash, A. J., Portfolio selection and skewness: evidence from international stock markets, *Journal of Banking and Finance*, 1997, Vol. 21, 143–167

26. Clarke, R. C., Krase, S. and Statman, M., Tracking errors, regret and tactical asset allocation, *Journal of Portfolio Management*, 1994, Vol. 20, 16–24

27. Cover, T. M., Universal portfolios, *Mathematical Finance*, 1991, Vol. 1, 1–29

28. Dantzig, G. B. and Infanger, G., Multi-stage stochastic linear programs for portfolio optimization, *Annals of Operations Research*, 1993, Vol. 45, 59–76

29. Deng, X. T., Li, Z. F. and Wang, S. Y., On computation of arbitrage for markets with friction, in Computing and Combinatorics, *Lecture Notes in Computer Science*, Berlin: Springer-Verlag, 2000, Vol. 1858, 310–319

30. Deng, X. T., Wang, S. Y. and Xia, Y. S., Criteria, models and strategies in portfolio selection, *Advanced Modeling and Optimization*, 2000, Vol. 2, 79–104

31. Deng, X. T., Li, Z. F. and Wang, S. Y., A minimax portfolio selection strategy with equilibrium, *European Journal of Operational Research*, 2005, Vol. 166, 278–292.

32. Dermody, J. C. and Rockafellar, R. T., Cash stream valuation in the face of transaction costs and taxes, *Mathematical Finance*, 1991, Vol. 1, 31–54

33. Dermody, J. C. and Rockafellar, R. T., Tax basis and nonlinearity in cash stream valuation, *Mathematical Finance*, 1995, Vol. 5, 97–119

34. Dong, J. C., Du, H. S., Wang, S. Y., Chen, K. and Deng, X. T., A framework of web-based decision support systems for portfolio selection with OLAP and PVM, *Decision Support Systems*, 2004, Vol. 37, 367–376

35. Dourra, H. and Siy, P., Stock evaluation using fuzzy logic, *International Journal of Theoretical and Applied Finance*, 2001, Vol. 4, 585–602

36. Dowd, K., *Beyond Value at Risk: The New Science of Risk Management*, London: John Wiley & Sons, 1998

37. Dubois, D. and Prade, H., *Possibility Theory*, New York: Plenum Press, 1988

38. Duffie, D. and Pan, J., An overview of value at risk, *Journal of Derivatives*, 1997, Vol. 4, 7–49

39. Duffie, D. and Sun, T., Transaction costs and portfolio choice in a discrete continuous time setting, *Journal of Economic Dynamics and Control*, 1990, Vol. 14, 35–51

40. Duffie, D. and Richardson, H.R., Mean-variance hedging in continuous time, *Annals of Applied Probability*, 1991, Vol. 1, 1–15

41. Dumas, B. and Luciano, E., An exact solution to a dynamic portfolio choice problem under transaction costs, *Journal of Finance*, 1991, Vol. 46, 577–595

42. Elton, E. J. and Gruber, M. J., The multi-period consumption investment problem and single period analysis, *Oxford Economics Papers*, 1974, Vol. 9, 289–301

43. Elton, E. J. and Gruber, M. J., On the optimality of some multiperiod portfolio selection criteria, *Journal of Business*, 1974, Vol. 7, 231–243

44. Fama, E. F., Multi-period consumption-investment decisions, *American Economic Review*, 1970, Vol. 60, 163–174

45. Fang, Y., Lai, K. K. and Wang, S. Y., Portfolio rebalancing with transaction costs and a minimal purchase unit, to appear in *Dynamics of Continuous, Discrete and Impulsive Systems* (B), 2005

46. Fang, Y., Lai, K. K. and Wang, S. Y., Portfolio selection models based on interval linear programming, in Chen, S., Wang, S. Y., Wu, Q. F. and Zhang, L. (eds.), *Financial Systems Engineering*, Lecture Notes in Decision Sciences, Hong Kong: Global-Link Publisher, 2003, Vol. 2, 19–38

47. Fang, Y., Lai, K. K. and Wang, S. Y., A fuzzy approach to portfolio rebalancing with transaction costs, in Sloot, P. M. A., Abramson, D., Dongarra, J. J., Bogdanov, A. V., Zomaya, A. Y. and Gorbachev, Y. E. (eds.), *Proceedings of Workshop on Computational Finance and Economics within International Conference on Computational Science*, Lecture Notes in Computer Science, Berlin: Springer-Verlag, 2003, Vol. 2658, 10–19

48. Fang, Y., Lai, K.K. and Wang, S.Y., Portfolio rebalancing models with transaction costs based on the fuzzy decision theory, *European Journal of Operational Research*, 2006, Vol. 175, Issue 2, 879–893

49. Fang, Y., Chen, L.H., Fukushima, M., A Mixed R&D Projects and Securities Portfolio Selection Model, appear in *European Journal of Operational Research*, 2007, doi:10.1016/j.ejor.2007.01.002.

50. Fang, Y. and Wang, S.Y., An interval semi-absolute deviation model for portfolio selection, *Lecture Notes in Computer Science*, 2006, Vol. 4223, 766–775

51. Fang, Y. and Wang, S.Y., A fuzzy index tracking portfolio selection model, *Lecture Notes in Computer Science*, 2005, Vol. 3516, 554–561

52. Fang, Y., Lai, K.K. and Wang, S.Y., A fuzzy mixed projects and securities portfolio selection model, *Lecture Notes in Computer Science*, 2005, Vol. 3614, 931–940

53. Fang, Y., Lai, K.K. and Wang, S.Y., Portfolio rebalance model with transaction costs, Proceedings of International Workshop on Strategic Management and Decision Analysis, Jouy-en-Josas, France, September 17-19, 2001, 176–190

54. Fernholz, R., Garvy, R. and Hannon, J., Diversity weighted indexing, *Journal of Portfolio Management*, Winter, 1998, 74–82

55. Fishburn, P. C., Mean-risk analysis with risk associated with below-target returns, *American Economic Review*, 1977, Vol. 67, 116–126

56. Ghezzi, L. L., A maxmin policy for bond management, *European Journal of Operational Research*, 1999, Vol. 114, 389–394

57. Guo, P., Tanaka, H., Possibility data analysis and its application to portfolio selection problems, *Fuzzy Economic Review*, 1998, Vol.3, 3–33

58. Guo, P., Tanaka, H., Decision analysis based on fused double exponential possibility distributions, *European Journal of Operational Research*, 2003, Vol. 148, 467–479

59. Hakansson, N. H., Multi-period mean-variance analysis: Toward a general theory of portfolio choice, *Journal of Finance*, 1971, Vol. 26, 857–884

60. Hamza, F. and Janssen, J., The mean-semivariances approach to realistic portfolio optimization subject to transaction costs, *Applied Stochastic Models Data Analysis*, 1998, Vol. 14, 275–283

61. Hansen, E., *Global Optimization Using Interval Analysis*, New York: Marcel Dekker, 1992

62. Heilpern, S., The expected value of a fuzzy number, *Fuzzy Sets and Systems*, 1992, Vol. 47, 81–86

63. Helmbold, D. P., Schapire, R. E., Singer, Y. and Warmuth, M. K., On-line portfolio selection using multiplicative updates, *Mathematical Finance*, 1998, Vol. 8, 325–347

64. Harlow, W. V., Asset allocation in a downside risk framework, *Financial Analysts Journal*, 1991, Vol. 47, 28–40

65. Hwang, C. L. and Yoom, K. S., *Multiple Attribute Decision Making: Methods and Applications*, New York: Springer-Verlag, 1981

66. Inuiguchi, M. and Ramik, J., Possibilistic linear programming: a brief review of fuzzy mathematical programming and a comparison with stochastic programming in portfolio selection problem, *Fuzzy Sets and Systems*, 2000, Vol. 111, 3–28

67. Inuiguchi, M. and Tanino, T., Portfolio selection under independent possibilistic information, *Fuzzy Sets and Systems*, 2000, Vol. 115, 83–92

68. Ishibuchi, H. and Tanaka, H., Formulation and analysis of linear programming problem with interval coefficients, *Journal of Japan Industrial Management Association*, 1989, Vol. 40, 320–329

69. Ishibuchi, H. and Tanaka, H., Multiobjective programming in optimization of the interval objective function, *European Journal of Operational Research*, 1990, Vol. 48, 219–225

70. Jacob, N. L., A limited diversification portfolio selection model for the small investor, *Journal of Finance*, 1974, Vol. 29, 847–856

71. Jansen, R. and Dijk, R. V., Optimal benchmark tracking with small portfolios, *Journal of Portfolio Management*, Winter, 2002, 33–39

72. Ji, X. D., Zhu, S. S., Wang, S. Y. and Zhang, S. Z., A dynamic stochastic programming model for multi-period portfolio selection, to appear in *IIE Transactions*, 2005

73. Jorion, P., *Value at Risk: The New Benchmark for Controlling Market Risk*, Chicago: Irwin, 1997

74. Kane, A., Skewness preference and portfolio choice, *Journal of Financial and Quantitative Analysis*, 1982, Vol. 17, 15–26

75. Kataoka, S., A stochastic programming model, *Econometrica*, 1963, Vol. 31, 181–196

76. Konno, H., Piecewise linear risk functions and portfolio optimization. *Journal of the Operations Research Society of Japan*, 1990, Vol. 33, No. 2, 139–156

77. Konno, K. and Yamazaki, H., Mean absolute deviation portfolio optimization model and its application to Tokyo stock market, *Management Science*, 1991, Vol. 37, 519–531

78. Konno, H. and Li, J., An internationally diversified investment using a stock-bond integrated portfolio model, *International Journal of Theoretical and Applied Finance*, 1998, Vol. 1, 145–160

79. Konno, H. and Kobayashi, H., An integrated stock-bond portfolio optimization model, *Journal of Economic Dynamics and Control*, 1997, Vol. 21, 1227–1244

80. Konno, H., Pliska, S. R. and Suzuki, K., Optimal portfolios with asymptotic criteria, *Annals of Operations Research*, 1993, Vol. 45, 187–204

81. Konno, H. and Shirakawa, H., Equilibrium relation in the mean-absolute deviation capital market, *Financial Engineering and the Japanese Market*, 1994, Vol. 1, 21–35

82. Konno, H., Shirakawa, H. and Yamazaki, H., A mean-absolute deviation-skewness portfolio optimization model, *Annals of Operations Research*, 1993, Vol. 45, 205–220

83. Konno, H. and Suzuki, K. A., Mean variance skewness optimization model, *Journal of the Operations Research Society of Japan*, 1995, Vol. 38, 173–187

84. Konno, H. and Wijayanayake, A., Mean absolute deviation portfolio optimization model under transaction costs, *Journal of the Operations Research Society of Japan*, 1999, Vol. 42, 422–435

85. Konno, H. and Wijayanayake, A., Portfolio optimization problem under concave transaction costs and minimal transaction unit constraints, *Mathematical Programming*, Ser. B, 2001, Vol. 89, 233–250

86. Konno, H. and Wijayanayake, A., Minimal cost index tracking under concave transaction costs, *International Journal of Theoretical and Applied Finance*, 2001, Vol. 4, 939–957

87. Konno, H. and Wijayanayake, A., Portfolio optimization under D.C. transaction costs and minimal transaction unit constraints, *Journal of Global Optimization*, 2002, Vol. 22, 137–154

88. Lai, K. K., Wang, S. Y., Zeng, J. H. and Zhu, S. S., Portfolio selection models with transaction costs: crisp case and interval number case, in Li, D.(eds.), *Proceedings of the 5th International Conference on Optimization Techniques and Applications*, Hong Kong, 2001, 943–950

89. Lai, K. K., Wang, S. Y., Xu, J. P., Zhu, S. S. and Fang Y., A class of linear interval programming problems and its application to portfolio selection, *IEEE Transactions on Fuzzy Systems*, 2002, Vol. 10, 698–704

90. León, T., Liern, V. and Vercher, E., Fuzzy mathematical programming for portfolio management, in Boilla, M. Casasus, T. and Sala, R.(eds.), *Financial Modelling*, Heidelberg: Physica-Verlag, 2000, 241–256

91. León, T., Liern, V. and Vercher, E., Viability of infeasible portfolio selection problems: a fuzzy approach, *European Journal of Operational Research*, 2002, Vol. 139, 178–189

92. Levy, H., Equilibrium in an imperfect market: a constraint on the number of securities in the portfolio, *American Economics Reviews*, 1978, Vol. 68, 643–658

93. Li, D. and Ng, W. K., Optimal dynamic portfolio selection: multi-period mean-variance formulation, *Mathematical Finance*, 2000, Vol. 10, 387–406

94. Li, R. J. and Lee, E. S., Ranking fuzzy numbers—a comparison, *Proceedings of North American Fuzzy Information Processing Society Workshop*, West Lafayette, IL, 1987, 169–204

95. Li, Z. F., Li, Z. X., Wang, S. Y. and Deng, X. T., Optimal portfolio selection of assets with transaction costs and no short sales, *International Journal of Systems Science*, 2001, Vol. 32, 352–365

96. Li, Z. F., Wang, S. Y. and Deng, X. T., A linear programming algorithm for optimal portfolio selection with transaction costs, *International Journal of Systems Science*, 2000, Vol. 31, 107–117

97. Lin, D. and Wang S. Y., A genetic algorithm for portfolio selection problems, *Advanced Modelling and Optimization*, 2002, Vol. 4, 13–27

98. Liu, B., *Theory and Practice of Uncertain Programming*, Heidelberg: Physica-Verlag, 2002

99. Liu, B., Minimax chance constrained programming models for fuzzy decision systems, *Information Sciences*, 1998, Vol. 112, 25–38

100. Liu, B. and Iwamura, K., A note on chance constrained programming with fuzzy coefficients, *Fuzzy Sets and Systems*, 1998, Vol. 100, 229–233

101. Liu, S., Wang, S. Y. and Qiu, W. H., Mean-variance-skewness model for portfolio selection with transaction costs, *International Journal of Systems Science*, 2003, Vol. 34, 255–262

102. Mao, J. C. T., Models of capital budgeting, E-V vs E-S, *Journal of Financial and Quantitative Analysis*, 1970, Vol. 5, 657–675

103. Mao, J. C. T., Essentials of portfolio diversification strategy, *Journal of Finance*, 1970, Vol. 25, No. 5, 1109–1121

104. Markowitz, H. M., Portfolio selection, *Journal of Finance*, 1952, Vol. 7, 77–91

105. Markowitz, H. M., The optimization of a quadratic function subject to linear constraints, *Naval Research Logistics Quarterly*, 1956, Vol. 3, 111–133

106. Markowitz, H. M., *Portfolio Selection: Efficient Diversification of Investment*, New York: John Wiley & Sons, 1959

107. Markowitz, H. M., *Mean-Variance Analysis in Portfolio Choice and Capital Markets*, Cambridge: Blackwell Publishers, 1987

108. Mansini, R. and Speranza, M. G., Heuristic algorithms for the portfolio selection problem with minimum transaction lots, *European Journal of Operational Research*, 1999, Vol. 114, 219–233

109. Merton, R. C., Lifetime portfolio selection under uncertainty: The continuous time case, *Review of Economics and Statistics*, 1969, Vol. 51, 247–257
110. Modigliani, F. and Miller, M. H., The cost of capital, corporation finance and the theory of investment, *American Economic Review*, 1958, Vol. 48, 261–297
111. Morton, A. J. S. and Pliska, R., Optimal portfolio management with transaction costs, *Mathematical Finance*, 1995, Vol. 5, No. 4, 337–356
112. Mossin, J., Optimal multi-period portfolio policies, *Journal of Business*, 1968, Vol. 41, 215–229
113. Mulvey, J. M. and Vladimirou, H., Stochastic network programming for financial planning problems, *Management Science*, 1992, Vol. 38, 1642–1664
114. Mulvey, J. M., Generating scenarios for the Towers Perrin investment system, *Interfaces*, 1996, Vol. 26, 1–15
115. Mulvey, J. M., Introduction to financial optimization: mathematical programming special issue, *Mathematical Programming*, Ser.B, 2001, Vol. 89, 205–216
116. Nakahara, Y., Sasaki, M. and Gen, M., On the linear programming with interval coefficients, *International Journal of Computers and Engineering*, 1992, Vol. 23, 301–304
117. Ogryczak, O. and Ruszczynski, A., From stochastic dominance mean-risk model: semi-deviation as risk measure, *European Journal of Operational Research*, 1999, Vol. 116, 33–50
118. Ogryczak, O. and Ruszczynski, A., On constancy of stochastic dominance and mean-semideviation models, *Mathematical Programming,* Ser. B, 2001, Vol. 89, 217–232
119. Östermark, R., Vector forecasting and dynamic portfolio selection: Empirical efficiency of recursive multiperiod strategies, *European Journal of Operational Research*, 1991, Vol. 55, 46–56
120. Östermark, R., A fuzzy control model (FCM) for dynamic portfolio management, *Fuzzy Sets and Systems*, 1996, Vol. 78, 243–254
121. Patel, N. R. and Subrahmanyam, M. G., A simple algorithm for optimal portfolio selection with fixed transaction costs, *Management Science*, 1982, Vol.28, 303–314
122. Parra, M. A., Terol, A. B. and Uría, M. V. R., A fuzzy goal programming approach to portfolio selection, *European Journal of Operational Research*, 2001, Vol. 133, 187–297
123. Perold,A. F, Large-scale portfolio optimization, *Management Science*, 1982, Vol. 28, 303–314
124. Perold, A. F. and Sharpe W. F., Dynamic strategies for asset allocation, *Financial Analysts Journal*, 1988, Vol. 44, 16–27
125. Peters, E. E., *Chaos and Orders in the Capital Markets*, New York: John Wiley & Sons, 1996
126. Philippe, J., *Value at Risk: The New Benchmark for Controlling Derivatives Risk*, New York: McGraw-Hill, 1997
127. Pogue, G. A., An existence of the Markowitz portfolio selection model to include variable transaction costs, short sales, leverage policies and taxes, *Journal of Finance*, 1986, Vol. 25, 1005–1028

128. Ramaswamy, S., Portfolio selection using fuzzy decision theory, Working Paper of Bank for International Settlements, 1998, No. 59

129. Rockafellar, R. T., *Convex Analysis*, Princeton: Princeton University Press, 1970

130. Rockafellar, R. T. and Uryasev, S., Optimization of conditional Value-at-Risk, *Journal of Risk*, 2000, Vol. 2, 21–41

131. Rockafellar, R. T. and Uryasev, S., Conditional Value-at-Risk for general loss distributions, *Journal of Banking and Finance*, 2002, Vol. 26, 1443–1471

132. Roll, R., A mean variance analysis of tracking error, *Journal of Portfolio Management*, Summer, 1992, Vol. 18, 13–22

133. Rommelfanger, H., Hanscheck, R. and Wolf, J., Linear programming with fuzzy objectives, *Fuzzy Sets and Systems*, 1989, Vol. 29, 31–48

134. Roy, A. D., Safety-first and the holding of assets, *Econometrics*, 1952, Vol. 20, 431–449

135. Ross, S. A., The arbitrage theory of capital asset pricing, *Journal of Economic Theory*, 1976, Vol. 13, 341–360

136. Ross, S. A., Risk, return and arbitrage, in *Friend, I. and Bicksler, J. L. (eds), Risk and Returns in Finance*, Cambridge: MIT Press, 1977, 189–218.

137. Rudolf, M., Wolter, H. J. and Zimmermann, H., A linear model for tracking error minimization, *Journal of Banking and Finance*, 1999, Vol. 23, 85–103

138. Rudolf, M., *Algorithms for Portfolio Optimization and Portfolio Insurance*, Bern: Paul Haupt Verlags AG, 1994

139. Saaty, T. L., *The Analytic Hierarchy Process*, New York: McGraw-Hill, 1980

140. Sadagopan, S. and Ravindran, A., Interactive solution of bi-criteria mathematical programs, *Naval Research Logistics Quarterly*, 1982, Vol. 29, 443–458

141. Samuelson, P., The fundamental approximation theorem of portfolio analysis in terms of means variances and higher moments, *Review of Economic Studies*, 1958, Vol. 25, 65–86

142. Sengupta, J. K., Portfolio decisions as games, *International Journal of Systems Science*, 1989, Vol. 20, 1323–1334

143. Sengupta, J. K., Mixed strategy and information theory in optimal portfolio choice, *International Journal of Systems Science*, 1989, Vol. 20, 215–227

144. Sharpe, W. F., Capital asset prices: a theory of market equilibrium under conditions of risk, *Journal of Finance*, 1964, Vol. 19, 425–442

145. Sharpe, W. F., A simplified model for portfolio analysis, *Management Science*, 1963, Vol. 9, 277–293

146. Sharpe, W., Alexander, G. F. and Bailey, J. V., *Investments*, London: Prentice Hall, 1995

147. Simaan, Y., Estimation risk in portfolio selection: the mean variance model versus the mean absolute deviation model, *Management Science*, 1997, Vol. 43, 1437–1446

148. Smith, J. E., Moment methods for decision analysis, *Management Science*, 1993, Vol. 39, 340–358

149. Speranza, M. G., Linear programming models for portfolio optimization, *Finance*, 1993, Vol. 14, 107–123

150. Steuer, R. E., Algorithm for linear programming problems with interval objective function coefficients, *Mathematics of Operational Research*, 1981, Vol. 6, 333–348

151. Sturm, J., Using SeDuMi 1.02, a MATLAB toolbox for optimization over symmetric cones, *Optimization Methods and Software*, 1999, Vol. 11, 625–653

152. Swalm, R. O., Utility theory-insights into risk taking, *Harvard Business Review*, 1966, Vol. 44, 123–136

153. Tanaka, H. and Guo, P., Portfolio selection based on upper and lower exponential possibility distributions, *European Journal of Operational Research*, 1999, Vol. 114, 115–126

154. Tanaka, H., Guo, P. and Türksen, I. B., Portfolio selection based on fuzzy probabilities and possibility distributions, *Fuzzy Sets and Systems*, 2000, Vol. 111, 387–397

155. Tanaka, H., Fuzzy data analysis by possibilistic linear models, *Fuzzy Sets and Systems*, 1987, Vol. 24, 363–375

156. Tanaka, H., Guo, P., Portfolio selection based on possibility theory. In: R.R. Ribeiro et al., Editors, Soft Computing in Financial Engineering, Physica-Verlag, Berlin, Heidelberg, 1999, 159–185

157. Tay, N. S. P. and Linn, S. C., Fuzzy inductive reasoning, expectation formation and the behavior of security prices, *Journal of Economic Dynamics and Control*, 2001, Vol. 25, 321–361

158. Telser, L. G., Safty first and hedging, *Review of Economic Studies*, 1955, Vol. 23, 1–16

159. Teo, K. L. and Yang, X. Q., Portfolio selection problem with minimax type risk function, *Annals of Operations Research*, 2001, Vol. 101, 333–349

160. Vandenberghe, L. and Boyd, S., Semidefinite programming, *SIAM Review*, 1996, Vol. 38, 49–95

161. Watada, J., Fuzzy portfolio model for decision making in investment, in Yoshida, Y.(eds.), *Dynamical Aspects in Fuzzy Decision Making*, Heidelberg: Physica-Verlag, 2001, 141–162

162. Wang, S. Y. and Xia, Y. S. *Portfolio Selection and Asset Pricing*, Berlin: Springer-Verlag, 2002

163. Wang, S. Y., Yamamoto Y. and Yu, M., A minimax rule for portfolio selection in frictional markets, *Mathematical Methods of Operations Research*, 2003, Vol. 57, 141–155

164. Wang, S. Y. and Zhu, S. S., On fuzzy portfolio selection problems, *Fuzzy Optimization and Decision Making*, 2002, Vol. 1, 361–377

165. Xia, Y. S., Liu, B. D., Wang, S. Y. and Lai, K. K., A model for portfolio selection with order of expected returns, *Computers and Operations Research*, 2000, Vol. 27, 409–422

166. Xia, Y. S., Wang, S. Y. and Deng, X. T., A compromise solution to mutual funds portfolio selection with transaction costs, *European Journal of Operational Research*, 2001, Vol. 134, 564–581

167. Yoshimoto, A., The mean-variance approach to portfolio optimization subject to transaction costs, *Journal of the Operational Research Society of Japan*, 1996, Vol. 39, 99–117

168. Young, M. R., A minimax portfolio selection rule with linear programming solution, *Management Science*, 1998, Vol. 44, 673–683

169. Yu, M., Dong, H. B., Wang, S. Y., *Portfolio Selection and Arbitrage in Fractional Financial Market*, (in Chinese), Beijing: Science Press, 2004

170. Yu, L. Y., Ji, X. D. and Wang, S. Y., Stochastic programming models in financial optimization: a survey, *Advanced Modelling and Optimization*, 2003, Vol. 5, 1–26

171. Yu, P. L., *Multiple Criteria Decision Making: Concepts Techniques and Extensions*, New York: Plenum Press, 1985

172. Zadeh, L. A., Fuzzy sets, *Information and Control*, 1965, Vol. 8, 338–353

173. Zadeh, L. A., Fuzzy sets as a basis for a theory of possibility, *Fuzzy Sets and Systems*, 1978, Vol. 1, 3–28

174. Zhang, S. M., Wang, S. Y. and Deng, X. T., Portfolio selection theory with different interest rates for borrowing and lending, *Journal of Global Optimization*, 2004, Vol. 28, 67–95

175. Zimmermann, H. J., *Fuzzy Set Theory and Its Applications*, Dordrecht: Kluwer Academic Publishers, 1985

176. Zhou, X. Y. and Li, D., Continuous-time mean-variance portfolio selection: a stochastic LQ framework, *Applied Mathematics and Optimization*, 2000, Vol. 42, 19–33

177. Zhu, S. S., Li, D. and Wang, S. Y., Risk control over bankruptcy in dynamic portfolio selection: a generalized mean-variance formulation, *IEEE Transactions on Automatic Control*, 2004, Vol. 49, 447–457

178. Zmeskal, Z., Application of the fuzzy-stochastic methodology to appraising the firm value as a European call option, *European Journal of Operational Research*, 2001, Vol. 135, 303–310

179. Zeng, J. H., Wang, S. Y. and Gao, J. W., A model for portfolio selection based on investment efficiency, *Advanced Modelling and Optimization*, 2002, Vol. 4, 111–121

180. Zeng, J. H., Wang, S. Y., Lai, K. K., Zhu, S. S., Portfolio selection models with transaction costs: crisp case and interval number case, In Proceedings of The 5th International Conference on Optimization: Techniques and Applications, Hong Kong, December 15-17, 2001, 943–950

Subject Index

Lecture Notes in Economics and Mathematical Systems

For information about Vols. 1–524
please contact your bookseller or Springer-Verlag

Vol. 569: L. Neubecker, Strategic Competition in Oligopolies with Fluctuating Demand. IX, 233 pages, 2006.

Vol. 570: J. Woo, The Political Economy of Fiscal Policy. X, 169 pages. 2006.

Vol. 571: T. Herwig, Market-Conform Valuation of Options. VIII, 104 pages. 2006.

Vol. 572: M. F. Jäkel, Pensionomics. XII, 316 pages. 2006

Vol. 573: J. Emami Namini, International Trade and Multinational Activity. X, 159 pages, 2006.

Vol. 574: R. Kleber, Dynamic Inventory Management in Reverse Logistics. XII, 181 pages, 2006.

Vol. 575: R. Hellermann, Capacity Options for Revenue Management. XV, 199 pages, 2006.

Vol. 576: J. Zajac, Economics Dynamics, Information and Equilibnum. X, 284 pages, 2006.

Vol. 577: K. Rudolph, Bargaining Power Effects in Financial Contracting. XVIII, 330 pages, 2006.

Vol. 578: J. Kühn, Optimal Risk-Return Trade-Offs of Commercial Banks. IX, 149 pages, 2006.

Vol. 579: D. Sondermann, Introduction to Stochastic Calculus for Finance. X, 136 pages, 2006.

Vol. 580: S. Seifert, Posted Price Offers in Internet Auction Markets. IX, 186 pages, 2006.

Vol. 581: K. Marti; Y. Ermoliev; M. Makowsk; G. Pflug (Eds.), Coping with Uncertainty. XIII, 330 pages, 2006.

Vol. 582: J. Andritzky, Sovereign Default Risks Valuation: Implications of Debt Crises and Bond Restructurings. VIII, 251 pages, 2006.

Vol. 583: I.V. Konnov, D.T. Luc, A.M. Rubinov† (Eds.), Generalized Convexity and Related Topics. IX, 469 pages, 2006.

Vol. 584: C. Bruun, Adances in Artificial Economics: The Economy as a Complex Dynamic System. XVI, 296 pages, 2006.

Vol. 585: R. Pope, J. Leitner, U. Leopold-Wildburger, The Knowledge Ahead Approach to Risk. XVI, 218 pages, 2007 (planned).

Vol. 586: B.Lebreton, Strategic Closed-Loop Supply Chain Management. X, 150 pages, 2007 (planned).

Vol. 587: P. N. Baecker, Real Options and Intellectual Property: Capital Budgeting Under Imperfect Patent Protection. X, 276 pages , 2007.

Vol. 588: D. Grundel, R. Murphey, P. Panos , O. Prokopyev (Eds.), Cooperative Systems: Control and Optimization. IX, 401 pages , 2007.

Vol. 589: M. Schwind, Dynamic Pricing and Automated Resource Allocation for Information Services: Reinforcement Learning and Combinatorial Auctions. XII, 293 pages , 2007.

Vol. 590: S. H. Oda, Developments on Experimental Economics: New Approaches to Solving Real-World Problems. XVI, 262 pages, 2007.

Vol. 591: M. Lehmann-Waffenschmidt, Economic Evolution and Equilibrium: Bridging the Gap. VIII, 272 pages, 2007.

Vol. 592: A. C.-L. Chian, Complex Systems Approach to Economic Dynamics. X, 95 pages, 2007.

Vol. 593: J. Rubart, The Employment Effects of Technological Change: Heterogenous Labor, Wage Inequality and Unemployment. XII, 209 pages, 2007.

Vol. 594: R. Hübner, Strategic Supply Chain Management in Process Industries: An Application to Specialty Chemicals Production Network Design. XII, 243 pages, 2007.

Vol. 595: H. Gimpel, Preferences in Negotiations: The Attachment Effect. XIV, 268 pages, 2007.

Vol. 596: M. Müller-Bungart, Revenue Management with Flexible Products: Models and Methods for the Broadcasting Industry. XXI, 297 pages, 2007.

Vol. 597: C. Barz, Risk-Averse Capacity Control in Revenue Management. XIV, 163 pages, 2007.

Vol. 598: A. Ule, Partner Choice and Cooperation in Networks: Theory and Experimental Evidence. Approx. 200 pages, 2007.

Vol. 599: A. Consiglio, Artificial Markets Modeling: Methods and Applications. XV, 277 pages, 2007.

Vol. 600: M. Hickman, P. Mirchandani, S. Voss (Eds.): Computer-Aided Scheduling of Public Transport. Approx. 424 pages, 2007.

Vol. 601: D. Radulescu, CGE Models and Capital Income Tax Reforms: The Case of a Dual Income Tax for Germany. XVI, 168 pages, 2007.

Vol. 602: N. Ehrentreich, Agent-Based Modeling: The Santa Fe Institute Artificial Stock Market Model Revisited. XVI, 225 pages, 2007.

Vol. 603: D. Briskorn, Sports Leagues Scheduling: Models, Combinatorial Properties, and Optimization Algorithms. XII, 164 pages, 2008.

Vol. 604: D. Brown, F. Kubler, Computational Aspects of General Equilibrium Theory: Refutable Theories of Value. XII, 202 pages, 2008.

Vol. 605: M. Puhle, Bond Portfolio Optimization. XIV, 137 pages, 2008.

Vol. 606: S. von Widekind, Evolution of Non-Expected Utility Preferences. X, 130 pages, 2008.

Vol. 607: M. Bouziane, Pricing Interest Rate Derivatives: A Fourier-Transform Based Approach. XII, 191 pages, 2008.

Vol. 608: P. Nicola, Experimenting with Dynamic Macromodels: Growth and Cycles. XIII, 241 pages, 2008.

Vol. 609: Y. Fang, K.K. Lai, S. Wang, Fuzzy Portfolio Optimization: Theory and Models. IX, 173 pages, 2008.

Vol. 610: M. Hillebrand, Pension Systems, Demographic Change, and the Stock Market. X, 176 pages, 2008 .

Vol. 611: R. Brosch, Portfolios of Real Options. XVI, 174 pages, 2008.

Vol. 612: D. Ardia, Financial Risk Management with Bayesian Estimation of GARCH Models. XII, 203 pages, 2008.